Ulrich Graf   Nancy van Schaik
Friedrich E. Würgler

Drosophila Genetics

A Practical Course

With 25 Figures

Springer-Verlag
Berlin Heidelberg New York
London Paris Tokyo
Hong Kong Barcelona
Budapest

Dr. sc. nat. Ulrich Graf
Prof. Dr. sc. nat. Friedrich E. Würgler
Institute of Toxikology
ETH and University of Zurich
Schorenstreet 16
CH-8603 Schwerzenbach, Switzerland

Prof. Ph. D. Nancy van Schaik
Department of Genetics
University of the Witwatersrand
1 Jan Smuts Ave., P.O. Wits
2050 Johannesburg, South Africa

ISBN-13: 978-3-540-54327-5   e-ISBN-13: 978-3-642-76805-7
DOI: 10.1007/978-3-642-76805-7

Library of Congress Cataloging-in-Publication Data. Graf, Ulrich. Drosophila genetics: a practical course / Ulrich Graf, Nancy van Schaik, Friedrich E. Würgler. p. cm. Includes bibliographical references and index.

1. Drosophila melanogaster – Genetics. I. Van Schaik, Nancy. II. Würgler, F.E. (Friedrich E.), 1936– . III. Title.
QH470.D7G73 1992   595.77′4 – dc20   91-43912 CIP

Production Editor: Herta Böning, Heidelberg

31/3145-5 4 3 2 1 0 – Printed on acid-free paper

Foreword

The Biological Sciences are in the midst of a scientific rev-
olution. During the past decade under the rubric of molecular biology, chemistry and physics have assumed an integral
role in biological research. This is especially true in genetics, where the cloning of genes and the manipulation of
genomic DNA have become in many organisms routine laboratory
procedures. These noteworthy advances, it must be empha-
sized, especially in molecular genetics, are not autonomous.
Rather, they have been accomplished with those organisms
whose formal genetics has been documented in great detail.
For the beginning student or the established investigator who
is interested in pursuing eukaryote molecular genetic research, _Drosophila_ _melanogaster_, with its rich body of formal
genetic information is one organism of choice. The book
"Drosophila Genetics. A Practical Course" is an indispensable source of information for the beginner in the biology
and formal genetics of _Drosophila_ _melanogaster_. The scope of
this guide, a revision and enlargement of the original German
language version, is broad and instructive. The information
included ranges from the simple, but necessary, details on
how to culture and manipulate Drosophila flies to a series of
more sophisticated genetic experiments. After completing the
experiments detailed in the text, all students - neophyte or
experienced - will be richly rewarded by having acquired a
broad base of classical genetics information relevant for the
biologist in its own right and prerequisite to Drosophila
genetics research - formal and/or molecular.

Davis, California,   Melvin M. Green
February 1991

Preface

The fruit fly <u>Drosophila</u> <u>melanogaster</u> is ideally suited to the practical demonstration of the basic phenomena of genetics. A large number of easily recognizable genetic markers, a generation time of only 10 days and simple culture methods make the fruit fly the eukaryote of choice for many geneticists. In this book we have compiled a number of experiments which illustrate various fields of genetics, ranging from simple Mendelian crosses to cytogenetic analyses and the induction of mutations in DNA repair defective mutants. We have deliberately limited the experiments to those which can be carried out without expensive equipment and have tried to describe the procedures in enough detail that teachers and students with little previous experience in Drosophila techniques can complete them successfully. Although primarily intended for university use, many of the experiments are simple enough to be used in a high school biology course. All have been tested in the class room. A Results and Answers section is included with the intention of illustrating the types of results obtained in real class situations and guiding students through scientific reasoning. In addition to references directly cited or furnishing detail or background information, we have referred to historic seminal papers in the hope that students will try to read at least some of the original work to develop a feeling for the history of genetics. Although based on the original German publication "Drosophila-Genetik; Praktikum der Genetik, Band 2" by F.E. Würgler and U. Graf, this book has been extensively modified. A section on the Molecular Biology of Drosophila has been added not to provide detailed experimental protocols but rather to give the student an overview for what is possible and to provide references to appropriate publications where techniques are given.

We are convinced that the technically simple Drosophila experiments described here ought to occupy a prime position in the teaching of genetics for beginners, and we hope that they will facilitate the experimental approach to the fascinating field of Drosophila genetics for both learners and teachers.

We wish to express our sincere thanks to Ruth Rümmele for her
expert typing and to our colleagues in the Genetics Group of
the Institute of Toxicology of the Swiss Federal Institute of
Technology and University of Zurich, Schwerzenbach, and in
the Department of Genetics of the University of the Witwatersrand, Johannesburg, for helpful discussions and sugges-
tions, although we must of course assume full responsibility
for any errors or omissions.

Schwerzenbach, Switzerland,   Ulrich Graf
April 1991   Nancy van Schaik
                                               Friedrich E. Würgler

Table of Contents

1. General

1.1 Why Drosophila?

<u>Drosophila</u> <u>melanogaster</u>, the fruit fly or vinegar fly, was
one of the first animals to be intensively studied genetically. In the laboratory of T.H. Morgan in the United States
soon after the rediscovery of Mendel's work, it was recognized as an ideal experimental animal for genetic studies
because of its small (but not too small) size, ease of culture, short generation time, large progeny size, low chromosome number, and giant salivary chromosomes. Morgan and his
students used Drosophila to elucidate the mechanisms of Mendelian inheritance and to construct the first linkage maps.

Cytogeneticists have found Drosophila a useful organism for
the study of chromosome morphology and karyotype evolution,
and population geneticists make use of quantitatively inherited characteristics for selection experiments and evolution-
ary studies. Complex traits such as circadian rhythm and
behavior can also be studied. As genetics moved from being a
descriptive science into a biochemical one and more lately a
molecular one, Drosophila proved useful for all types of
analysis. It remains a favorite experimental animal for mutation study and genetic toxicology. It is one of the first
eukaryotic organisms on which genetic engineering can be carried out relatively easily and in which the molecular basis
of development can be studied.

As an organism for teaching genetics, it has the advantage
that its life cycle fits conveniently into weekly laboratory
schedules. The techniques required for simple experiments
can be mastered by students with little scientific background, and most importantly, the fruit fly is still one of
the cheapest and least demanding of all laboratory organisms.

1.2 Basic Equipment

Each student will need the following:
<u>Stereo dissecting microscope</u> capable of magnifications up to
25x. Zoom type continuous magnification change is convenient

but not necessary. A strong and easily adjustable light source is important. If available, a heat filter for the lamp will help to avoid dehydrating live flies during examination.

<u>Compound microscope</u> with magnification up to at least 100x and an oil immersion objective (for the examination of salivary gland preparations).

<u>Anesthetizer</u>. This is basically a small bottle with some way of administering ether (for details see section 1.5).

<u>Morgue</u> for the discarding of flies no longer needed. A beaker or can filled with oil (used motor oil works very well) or alcohol (70%) is convenient.

<u>Tapping pad</u>. A solid rubber, foam rubber or felt pad (about 6 x 10 cm, 1 cm thick) for tapping flies from one container into another helps prevent breakage and cuts down on noise.

<u>White tile</u>. Flies can be moved around easily during examination or sorting on a slick white surface. Ordinary white bathroom tiles work very well, or a thick glass plate backed with white paper can be used.

<u>Brush</u> to move flies without damaging them. A fine camel's hair brush is best but a cheap number 1 or 2 watercolor brush can be used for student work.

<u>Forceps</u> for the preparation of larvae and handling individual flies. These should have fine pointed tips.

<u>Storage</u>. It is convenient for each student to have a box or drawer in which his own small items of equipment can be kept.

1.3 Genetic Terminology

<u>Allele</u>. One of a series of possible alternative forms of a given gene unique in DNA sequence and affecting the functioning of a single product (RNA and/or protein). If more than two mutant alleles have been identified in a species, the locus is said to show "multiple allelism".

<u>Autosome</u>. A chromosome other than a sex chromosome.

<u>Chromosome mutation (chromosomal aberration)</u>. A change affecting more than one gene such as the loss, duplication, or rearrangement of chromosome material containing genetic information.

<u>Diploid condition</u>. The chromosome state, in which each type of chromosome (except the sex chromosomes) is represented twice in somatic cells (2N).

<u>Dominant/recessive</u>. In diploid organisms an allele which is phenotypically expressed in either the homozygous or heterozygous state is said to be dominant. An allele which is expressed only when homozygous is said to be recessive.

F_1. First filial generation; the offspring from an experimental crossing. The parents of the F_1 are referred to as P.

F_2. The progeny produced by intercrossing F_1 individuals.

<u>Gene</u>. A gene is an hereditary unit, in the classical sense, and usually occupies a specific position (locus) on a specific chromosome within the genome; operationally it is a unit that has one or more specific effects on the phenotype of the organism; a gene can mutate to various allelic forms and recombines with other such units; at the molecular level a specific sequence of nucleotides in a DNA molecule.

<u>Genome mutation</u>. A change in the number of complete chromosomes, giving rise to heteroploid cells or individuals with a chromosome number different from the normal for the species. Such a change may involve entire chromosome sets (euploidy) or one or more individual chromosomes (aneuploidy).

<u>Genotype</u>. The genetic constitution of a cell or of an organism, as distinguished from its physical appearance (phenotype).

<u>Haploid number</u>. The gametic chromosome number, symbolized by N.

<u>Hemizygous</u>. Having a gene present in a single dose. May refer to a gene in a haploid organism, or to a sex-linked gene in the heterogametic sex, or to a gene in the appropriate chromosomal segment of a deficiency heterozygote.

<u>Heterogametic sex</u>. The sex that produces gametes nonidentical for sex chromosomes (e.g. males of Drosophila and of mammals produce X- and Y-bearing sperm, usually in equal proportions).

<u>Heterozygous</u>. Having different alleles at the corresponding loci of homologous chromosomes.

<u>Homogametic sex</u>. The sex that normally produces gametes all of which carry only one kind of sex chromosome, e.g. the eggs

of female mammals and female Drosophila all carry an X chromosome.

<u>Homologous chromosomes</u>. Chromosomes that pair during meiosis and usually carry the same set of genes. Each homolog is a duplicate of one of the chromosomes contributed at syngamy by the mother or the father. Homologous chromosomes contain the same linear sequence of genes and as a consequence in a diploid organism each gene is present in duplicate.

<u>Homozygous</u>. Having identical rather than different alleles at the corresponding loci of homologous chromosomes and, therefore, breeding true.

<u>Interchromosomal aberrations</u> are those such as translocations in which pieces from more than one chromosome are involved.

<u>Intrachromosomal aberrations</u> involve changes within one chromosome. In addition to inversions, these include <u>deficiencies</u> and <u>duplications</u> which result in a change in the amount of genetic material in the chromosome.

<u>Inversions</u> and <u>shifts</u> involve changes in the arrangement of genetic material, but not in its amount. In the case of an inversion a chromosomal segment has been excised, turned around 180°, and reinserted at the same position on a chromosome, with the result that the gene sequence for the segment is reversed with respect to that of the rest of the chromosome. In the case of shift, a chromosomal segment has been removed from its normal position and inserted (in the normal or reversed sequence) into another region of the same chromosome.

<u>Lethal mutation</u>. A mutation which results in the premature death of the cell or organism carrying it. Dominant lethals kill heterozygotes, whereas recessive lethals kill only homozygotes.

<u>Locus</u>. (plural: loci) The position that a gene occupies on a chromosome.

<u>Mutation</u>. (1) The process by which a gene undergoes a chemical or structural change; (2) a modified gene resulting from mutation; (3) by extension, an individual manifesting the mutation.

<u>Phenotype</u>. The observable properties of an organism, produced by the genotype in conjunction with the environment.

<u>Point (gene) mutation</u>. Originally, a mutation affecting only

one gene (cf. chromosome mutation). As defined in molecular
genetics, a mutation caused by the substitution of one nucleotide for another (base pair substitution).
<u>Polyploid</u>. An individual or cell having more than two sets
of chromosomes.
<u>Reciprocal crosses</u>. Crosses of the forms A ♀ x B ♂ and
B ♀ x A ♂.
<u>Sex chromosomes</u>. The chromosomes that are dissimilar in the
heterogametic sex. In Drosophila called X and Y.
<u>Sex-linked</u>. Refers to genes located on the X chromosome.
<u>Test cross</u>. A cross between a heterozygote or individual of
unknown genotype to an individual homozygous for the recessive alleles of all genes involved. Useful for determining
genotypes, linkage relationships, etc.
/ (a slash) is used in genetic formulas to separate the symbols of alleles present on two homologous chromosomes.
; (a semicolon) separates symbols of genes located on nonho-
mologous chromosomes (e.g. se^+/se and y; st).

References

KING, R.C., STANSFIELD, W.D.: A Dictionary of Genetics. 3rd
    ed. New York, Oxford: Oxford University Press 1985.
RIEGER, R., MICHAELIS, A., GREEN, M.M.: Glossary of Genetics
    and Cytogenetics, Classical and Molecular. 4th ed.
    Berlin, Heidelberg, New York: Springer 1976. (New
    edition in preparation).

The use of symbols for genes and alleles

With the methods of classical genetics, the genetic determination of a trait can only be proven after a mutation in the
gene controlling this characteristic has been found. The
mutant allele is identified based on its expression as a different or "mutant" phenotype. Dominant alleles are immediately detectable, recessive ones only after they have become
homozygous (the hemizygous situation is an exception to this
rule).
Since the early days of Mendelian genetics it has been the
custom to name the gene after the phenotypic change observed
in the mutant. For example in Drosophila, a fly was found

with wings of reduced size. This phenotype was called "vestigial", and vg is used as the symbol. Symbols with the first letter in upper case denote a dominant mutation, those with a lower case first letter a recessive mutation. The original allele present before the mutation occurred is said to be "normal" or "wild type". The wild type allele is the most common allele found in natural populations. For practical reasons the name used for the mutant phenotype is also used to designate the mutant allele and the gene as such. In Drosophila the following notations are used: The new allele originating from a mutation is: vg. The wild type allele is the same symbol with a plus superscript (vg^+) often abbreviated simply +. The name, in this case "vestigial", is also used for the gene of which we now have two alleles vg^+ and vg. The "vestigial" gene controls the size of the wings. Note that the symbolism system used in Drosophila is different from that used for many other organisms in that the upper and lower case of the same letter, e.g. B and b indicate different genes and not alleles. B is a dominant mutation (Bar eye) and its wild type allele is B^+, while b is a recessive mutation (black body color) with normal body color b^+. Unfortunately it is not possible to restrict the designation of alleles to a simple two-letter code. Often several alleles of a gene are known (multiple alleles). They need more complicated symbols (e.g. $mei-9^{L1}$). A list of all the symbols used by Drosophila geneticists (and a description of the different phenotypes) is found in the book "Genetic Variations of <u>Drosophila melanogaster</u>" by Lindsley and Grell (1968).

Another convention is used to describe the localization of a given gene within the genome of Drosophila. To describe the position of the vg gene the following notation is used: vg (2-67.0). The first number within the parentheses designates the chromosome on which the gene is located (1 = X chromosome; 2, 3, 4 = autosomes). The second number after the dash describes the position of the gene on the chromosome. This number refers to map units determined in recombination experiments (see experiments 3.3 and 3.6). By convention, position 0.0 is on the end of the left (or short) arm of the chromosome.

To perform an experimental cross with Drosophila, female and male flies are placed together in a vial or bottle. Gene symbols are used to label bottles with the crosses made. As a rule the genotype of the female parent is written first, followed by an x and then the genotype of the male. If the genes involved in the cross are autosomal, the genetic formulas can be written without special symbols for the sex chromosomes. If we wished to indicate that homozygous vestigial winged females had been crossed to homozygous normal winged males, we would write: $vg/vg \times vg^+/vg^+$. In cases where we are interested in several genes on a chromosome, the symbols of the genes on one homolog are written together and separated from those on the other homolog by two horizontal lines, e.g.:

$$\frac{vg^+ \quad bw^+}{vg \quad bw}.$$

A general rule that is used in genetics is that only the alleles of interest are written although every fly is of course carrying two alleles of each of thousands of genes. Any gene not written is assumed to be homozygous for the wild type (+) allele.

Genetic symbols are also used to write genotypes in order to determine expected phenotypes and expected progeny from a particular cross. The genotype of an individual fly is written as if it consisted of only one cell, the zygote (or fertilized egg) from which the adult individual developed. To predict the results from a cross, the different kinds of gametes that can be formed by each parent must be worked out and then each egg genotype combined with each sperm genotype. For simple crosses this can be done by inspection, but for more complex crosses a genetic checkerboard diagram is used (see chapter 3).

Let us look at the expected results from the cross of vestigial winged females with wild type males:

<u>Genotypes</u>:
♀ vg/vg: This female forms only one type of egg: vg.
♂ vg^+/vg^+: This male forms only one type of sperm: vg^+.
The only progeny genotype expected is: vg^+/vg.

Note that by convention in a heterozygote, the dominant allele is usually written first.

What would you expect the phenotype of these flies to be?
Expected phenotype : ....

Note: Answers to questions are found in chapter 9. Play fair and don't consult the answers until you have worked out your own answers!

1.4 The Biology of *Drosophila melanogaster*

<u>Drosophila</u> <u>melanogaster</u> is an outstanding experimental organism for classroom experiments. A number of factors make Drosophila a favorite "guinea pig" of geneticists and recently also of developmental biologists and molecular biologists. A general overview of its biology will help you to become more familiar with this organism.

<u>The life cycle</u>: <u>Drosophila</u> <u>melanogaster</u> as a fly belongs to the holometabolic insects which undergo a complete metamorphosis. The sequence and the approximate duration of the different stages of the life cycle at its optimum growth temperature, 25°C, are:

Embryonic development	1 day
First larval instar (L1)	1 day
Second larval instar (L2)	1 day
Third larval instar (L3)	2 days
Prepupa	4 hours
Pupa	4.5 days
Adult stage (imago)	40-50 days

Males become sexually mature and fertile shortly after emergence from the pupal case. Females mature less quickly: 6 to 12 hours depending on the strain. Once a female is sexually mature she will mate repeatedly with different males. Sperm is stored by the female in the ventral receptacle and used to fertilize eggs laid at later times. For this reason

it is necessary to use virgin females in any cross in which
the males to be used are different from the ones in the culture bottle from which the females are taken. (See section
1.5 on the collection of virgin females).
Although the mean life span for adults at 25°C ranges from 40
to 50 days and individuals may live as long as 80 days, young
flies, not more than 10 days old, are normally used in
crosses since fertility decreases with age.
Under standard conditions (25°C and 60% relative humidity)
the whole development from egg to adult takes about 10 days.
At lower temperatures, the life cycle is lengthened: 14 days
at 23.5°C; 21 days at 18°C. It is convenient to keep stocks
not being used currently at 18°C so as to reduce the labor
involved in routine transfer of cultures. Development can be
speeded up by increasing temperature, but at around 29-30°C
pupal lethality increases drastically, and in most strains,
females become sterile at temperatures above 30°C.

Number of progeny: From a single pair of flies, kept for
several days in a culture vial, a progeny of several hundred
may be obtained. With Drosophila it is possible to obtain a
large progeny from a few parental flies within approximately
2 weeks. This is a prerequisite for experiments which need
statistical analyses.

Size: Adult flies as well as all the developmental stages
are fairly small. An adult fly is about 5 mm long and has a
weight of approximately 1 milligram. Thus, large populations of hundreds of individuals can be grown in relatively
small culture bottles. If cultures become overcrowded, the
size of individual flies may be markedly reduced.

Culture: The materials needed are fairly inexpensive (for
details see section 1.5). The preparation of the culture
medium for vials or bottles is very simple. Glassware can be
washed and reused.

Phenotypes: On adult flies, many genetically controlled
structures and colors are visible with the naked eye and many
other characteristics are easily detected under a low power
stereomicroscope by the inspection of anesthetized flies.

For classroom experiments, mutants with changes in body col-
or, eye color, wing size, shape of the eyes, etc. are easy to
see. In research, in addition to these morphological markers, biochemical differences (such as enzyme variants) and
behavioral characteristics (e.g. geotaxis, neurological
changes), etc. are also studied.

<u>Genetic organization</u>: For many experiments, it is of great
advantage that all the genetic material of the haploid set is
present in only four chromosomes. The haploid chromosome
number is x=4, and the somatic number (diploid) 2N=2x=8. The
number of chromosomes in the haploid set is identical with
the experimentally determined number of linkage groups (see
experiment 3.5). The small chromosome number makes it easy
to determine the linkage group of a newly discovered mutant.
The difference in sex chromosome constitution between the
homogametic females (XX) and the heterogametic males (XY) is
of special importance. In males, the X chromosome is present
only once and therefore genes on the X are present in a hemizygous condition. One consequence of this is that mutations
of genes on the X chromosome (sex-linked genes) are phenotypically expressed in males irrespective of their dominance;
recessive as well as dominant mutations are expressed in
males. Therefore, by choosing sex-linked genetic markers,
one can often save time in experiments.

<u>Giant chromosomes</u>: In some tissues of the larvae, e.g. in
the salivary gland and the Malpighian tubules, polytenic
giant chromosomes are present. A microscopic study of these
special chromosomes is particularly easy if third instar larvae are used.
Giant chromosomes in the salivary glands are formed by somatic pairing of homologous chromosomes and a 9- to 10-fold duplication of the chromatids resulting in polytene chromosomes
each made up of about 1000 strands within the cell nucleus.
Since these chromosomes are in interphase, they are much more
extended in length than chromosomes examined during mitosis.
Compared to chromosomes of normal somatic cells, giant chromosomes show a characteristic banding pattern. A detailed
knowledge of the banding pattern allows the identification of
small chromosomal regions, for example the precise position

of break points of chromosome aberrations. Because only homologous regions of chromosomes pair, the giant chromosomes of individuals heterozygous for chromosome aberrations show characteristic pairing structures such as gaps, loops or other pairing figures. In this way preparations of salivary gland chromosomes also demonstrate the consequences of homologous pairing similar to that occurring in meiosis.

The exact analysis of chromosome aberrations (in particular of deletions) in combination with the study of the Mendelian segregation of gene mutations allows precise cytogenetic localization of genes within the banding pattern of a particular chromosome.

In recent years, giant chromosomes have also been used for microsurgical experiments, e.g. individual bands have been isolated and their DNA cloned!

Genetic control of meiosis: In Drosophila melanogaster, a very special genetic control of meiosis is encountered. During oogenesis in females, crossing over occurs regularly during meiosis. In contrast to this, in spermatogenesis in males, there is no meiotic crossing over. This peculiarity of Drosophila melanogaster is very useful, particularly in the construction of new genetically marked stocks. The details of the genetic control of the meiotic events in Drosophila melanogaster can now be investigated thanks to the discovery of mutations affecting crossing over in females (meiotic mutants).

Stocks: The mode of inheritance of many mutant characteristics of adults and their preceding developmental stages have been studied. The dominance relationship is known for a number of alleles; the linkage group and the localization on the chromosome are known; often even the localization of the change within a gene producing a particular allele, and the localization with respect to the banding pattern of giant chromosomes is known. Stocks carrying these mutations are kept in research laboratories and in special "stock centers" from where they can be obtained. For the availability of the stocks used for the experiments described in this book, see the stocklist in section 1.6.

So far, we have described the experimental advantages of <u>Dro-sophila</u> <u>melanogaster</u> for the study of Mendelian genetics. Today this "domesticated animal of the geneticists" has a much wider use and importance in biological research. All of the modern applications take advantage of the extensive knowledge of the genetics and cytogenetics of the organism. New possibilities became available through the development of synthetic media for larvae and flies as well as for insect organ and cell culture. Some very specialized techniques have been developed; for example the transplantation of imaginal discs, the separation and reaggregation of disc cells, the transplantation of cell nuclei as well as biochemical micromethods. In this way, suitable tools have become available for the study of the genetics, physiology and biochemistry of whole animals, isolated organs or even individual cells. For many problems in developmental biology, unique experimental approaches have become possible. Outstanding examples are the phenomenon of transdetermination and the studies of homeotic mutants. Important contributions to the understanding of basic biological phenomena have resulted from the application of modern molecular biology technologies such as DNA cloning, sequencing and transformation in Drosophila research.

1.5 Methods for Culturing and Studying
Drosophila melanogaster

<u>Culture medium</u>

Different media are used in different research laboratories. Basically, the majority of the media contain water, sugar, agar-agar, corn (maize) meal, yeast, a bactericidal agent and a mold suppressor.
The preparation of the medium we use for research and classroom applications is described below. To make the preparation as easy as possible and avoid the inconvenience of weighing, all quantities are given as volumes in milliliters (ml). It is convenient to have a specially marked container for each ingredient.

In an ordinary cooking pot, mix together:

    3 l water

   45 ml powdered agar-agar

  150 ml fresh yeast

  260 ml sucrose.

Heat to boiling and then stir in:

  550 ml corn (maize) meal.

Heat again until mixture comes to the boil. Keep boiling gently for approx. 30 min while stirring. Remove from heat and allow to cool to 50 to 60°C. Stir in:

  10 ml Nipasol-M-sodium (trade name for

      propyl-p-hydroxy-benzoate)

  10 ml of an aqueous 0.1 % solution of streptomycin.

or: 17 ml of a solution of 10 parts propionic acid and

      1 part phosphoric acid.

Pour while still hot into culture bottles to make a layer about 3 cm thick (makes approx. 100 bottles).

We find it convenient to let the open bottles dry under a gauze net (to keep flies out) for 2-3 hours. The drying can be speeded up by blowing air over the bottles with an oscillating fan. The drying should not be too vigorous as medium that has become too dry will come loose from the bottles when they are inverted to shake out flies.

The surface of the cooled medium is then seeded with live baker's yeast. Mix

  125 ml live yeast with

  25-40 ml water to form a thick pap.

Using a thick pipette 2-3 drops are dropped onto the surface of the culture bottle (1 drop into a vial). An alternative method is to sprinkle the surface with lyophyllized baker's yeast. The culture vessels are then stoppered with foam rubber stoppers or cotton bungs.

Freshly prepared vials and bottles may be used for starting cultures within the next few days. If they are to be kept for a longer time, they should be stored in a refrigerator before live yeast is added. Cold bottles should be warmed up to room temperature, yeasted and checked for condensation before use.

Yeast, corn meal and sugar are obtainable from local grocery stores. Agar-agar, antibiotics and fungicides may be obtained from pharmaceutical suppliers. Two of the most useful fungicides are Methyl-Parasept (Teneco Chemicals Inc., 290 River Drive, Garfield, NJ 07026 USA) and Nipasol (Nipa Laboratories Ltd., Treforest Industrial Estate, Pontypridd, Glamorgan, South Wales, Great Britain).

Culture vessels

For rearing mass cultures, glass bottles with a volume of about 200 ml are useful. In many countries, bottles used for selling milk, cream or yoghurt are suitable for Drosophila cultures. If the bottles are of colorless glass, it is especially easy to follow the development of the culture through the walls, but colored bottles may also be used. About 40 ml of medium is used in a 200 ml bottle.

For the progeny of single pair matings (or the progeny from a small number of parental flies) glass or clear plastic vials with a volume up to 40 ml may be used. If experiments on a larger scale are planned, a careful analysis of the availability and cost of vials on the local commercial market is advisable. In vials of about 40 ml total volume, containing about 10 ml medium, progenies of up to 200 individuals may be produced. The handling of vials and bottles is facilitated if groups of these can be moved around. Wire baskets are ideal so that air can easily circulate around the bottles or vials in the culture room or incubator. In our laboratory, we use baskets made of plastic coated wire. The large baskets hold 24 bottles or 85 vials. Smaller ones which can be fitted into commercially available incubators are also convenient. If one chooses reusable glass vials and bottles, they should be washed with a good laboratory detergent as soon as possible after use, or else autoclaved, if they have to be kept for some days before washing.

Handling of adult flies

In order to examine the flies of various phenotypes used to start new crosses, the live flies have to be anesthetized. Two different gases suitable for this purpose are ether and carbon dioxide. In classroom experiments, ether is usually used because the equipment is much simpler than that required for carbon dioxide.

> **WARNING:**
> Mixtures of ether and air may explode: The boiling
> point of ether is about 35°C. Ether is a strong nar-
> cotic. Enough fresh air is of primary importance at
> the working place. Ether has to be kept in bottles
> made of brown glass to avoid the production of dan-
> gerous peroxides. No smoking or use of any open
> flame or device producing sparks is allowed in the
> laboratory.

The basic principle for the use of ether is to introduce the
flies for a short while into a container in which the flies
are exposed to ether vapor. After this anesthetic treatment,
the flies can be brought back to normal air and can be inspected under the stereomicroscope.
Sometimes, e.g. if progeny have to be classified according to
several phenotypes, quite some time is needed to inspect the
anesthetized flies. In this case, the duration of the anesthesia may be too short. The flies have to be reanesthetized
(without changing their position on the glass plate) before
they start waking up.

> **NOTE:**
> An anesthetizer must be examined before use to make
> sure that it is not contaminated with flies or eggs.

One type of anesthesia device (Figure 1) consists of a 2 dl
glass bottle (1), e.g. milk bottle. The bottom of the bottle
is covered with loose wadding or absorbent cotton (2). From
an ether bottle (about 40 ml) a few drops of ether are pi-
petted on the wadding. The anesthesia bottle is closed with
a cork stopper (5). A plastic or metal funnel (3) is inserted into the center of this stopper. The funnel is connected to a small vessel (4), made of ether-resistant material. Ether diffuses into the vessel through small holes in
the wall and anesthetizes the flies in the vessel. The di-
mensions of the funnel are of special importance because the
top openings of the culture bottles as well as those of the

vials have to fit snugly into the funnel to shake flies into
the anesthetizing chamber.

A second type of anesthetizer, especially suitable for classroom work, is shown in Figure 2. It consists of a glass bottle (1), a funnel (2) and a stopper (3). With this model the
flies are put directly into the glass bottle. Ether evapo-
rating from a wadding ring in the stopper fills the whole
bottle with ether vapor. The ring (4) in the stopper is easily cut out with a knife, filled with absorbent cotton and
held in place with a thin string. Another possibility is to
fix a pad of absorbent material (5) with glue or pin to the
base of the stopper. To use this device, the cotton pad is
dampened with a little ether and the stopper then replaced in
the bottle so that the ether fumes diffuse through the
bottle. Care should be taken that no liquid ether falls into

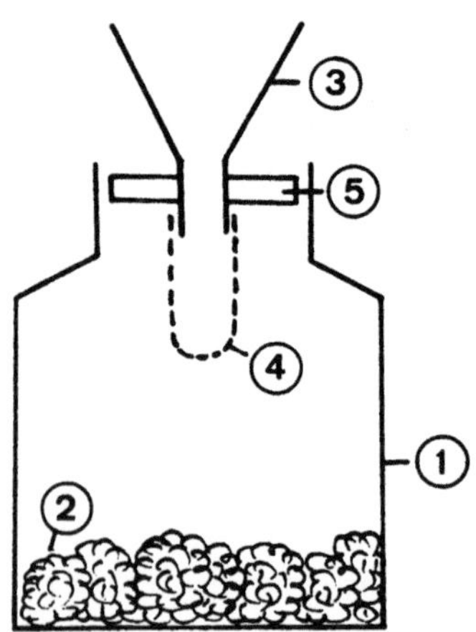

Figure 1. Anesthetizer I.

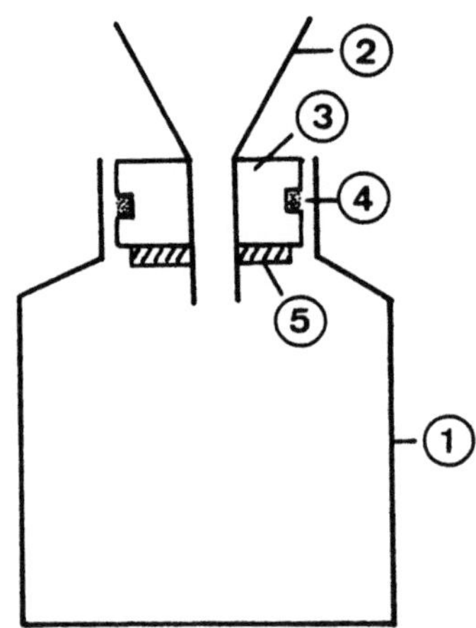

Figure 2. Anesthetizer II.

the bottle, because flies coming in direct contact with liquid ether die.

Flies in the culture vessels are shaken down to the medium by
tapping the bottle on a rubber pad. The stopper is now removed, the bottle turned upside down and fitted into the funnel. The flies are then shaken through the funnel into the
bottle. At first the flies will walk around on the bottom

and on the sides of the glass, but soon one can see that
their movements slow down. Finally they fall to the bottom
of the bottle. After all flies are inactivated, they are
left for a further 10-20 seconds in the ether in order to
extend the period of time that the flies will remain anesthetized. If the flies are needed later for setting up fresh
cultures, the anesthetic period has to be controlled carefully, because longer ether exposure will kill them. Flies dy-
ing from overetherization first reflex their wings above the
thorax and fold their legs. Eventually the legs are
stretched out. These symptoms are the consequences of excessive muscle contractions. When using any of the described
ether anesthesia devices, if any of the flies start to show
these symptoms, they must be put into fresh air immediately
or they will die.
For the inspection of the external characteristics, the flies
are shaken out of the anesthetizer and put on a glass plate
backed with white paper or on a white ceramic tile (size
about 8 x 12 cm). Under the stereomicroscope the flies can
best be studied using magnifications between 6 and 25x.
It is easy to see when the flies are starting to revive from
anesthesia; they start first to vibrate their legs and then
to stand up. After they have been standing for a short
while, they start to move around and eventually fly away.
Figure 3 shows a simple device that allows the reanesthetizing of the awaking flies. One takes one half of a glass Pet-
ri dish (1) (plastic dishes are unsuitable, because they are
often not ether-resistant) and fixes a piece of ether-resistant foam rubber in the center of the dish with an ether-resistant glue (2). In order to reanesthetize the flies, the
foam rubber is dampened with ether and the device is put over

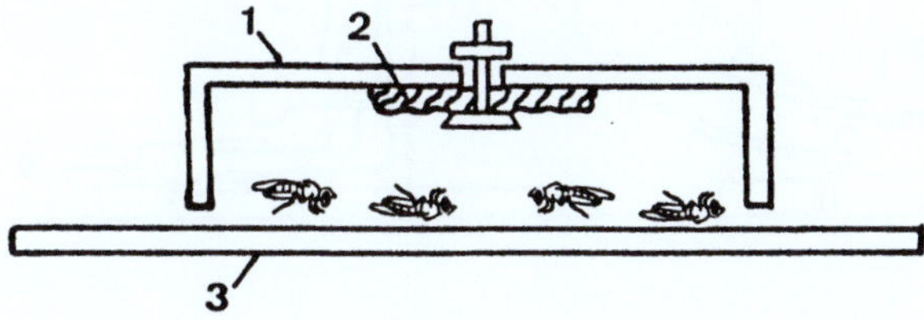

<u>Figure 3</u>. Glass Petri dish for reanesthesia.

the flies on the glass plate (3). Through the glass it is
easy to follow the reanesthesia, and after a relatively short
period, the device can be removed and the work with the flies
continued.
The risk of killing flies by overanesthesia is much lower if
carbon dioxide is used instead of ether. Technical problems
usually prevent the use of carbon dioxide in the classroom at
every working place, but for teachers and assistants working
regularly with the preparation of experiments, the use of
carbon dioxide has many advantages. For this reason a set-up
we have used for over 20 years is described.
Figure 4 shows a diagram of the device. A Bunsen burner
equiped for a pilot flame (1) is connected by plastic tubing
to a CO_2 cylinder (2). The air inlet (3) of the burner is
sealed with glue. When the valve of the burner (4) is in the

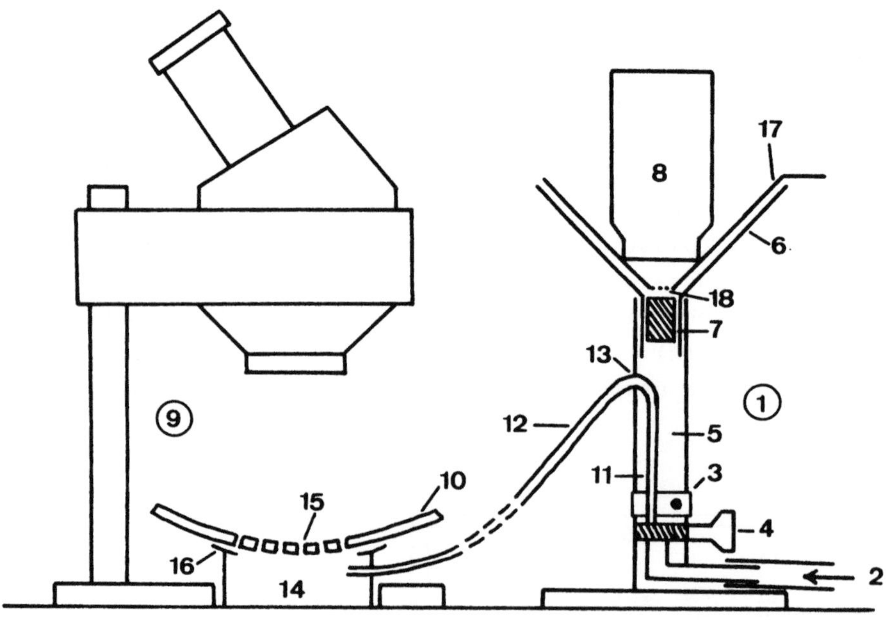

__Figure 4__. Equipment for CO_2 anesthesia.

"flame" position, the carbon dioxide flows through the large
tube of the burner (5) and enters a plastic funnel (6) fixed
to the top of the burner. A foam rubber stopper (7), through
which the gas can penetrate, is placed in the neck of the
funnel. A second, loose funnel (17) with its opening covered
with fine gauze (18) is fitted into the first one. Bottles
or vials (8) containing flies are put upside down into the
funnel. Upon gentle shaking the flies fall into the funnel
and are immediately immobilized by the CO_2 and accumulate
on the gauze. For inspection of the flies under a microscope
(9) the loose funnel is removed with the flies. They are
shaken into a concave plastic dish (10). In order to keep
the flies continuously in a CO_2 atmosphere, the valve (4)
is switched to the "pilot" position. Now the gas stream
passes through the thin tube (11) of the burner. On top of
the shortened thin tube a plastic tubing (12) is fixed. This
plastic tube passes through a hole (13) in the wall of the
large tube (5) and is connected to a cylindrical box (14).
From this box the gas stream reaches the flies through a
large number of very small holes (15) in the plastic dish.
This dish is made by cutting out a circular piece from a
conventional plastic bowl. The edge of the box (14) is
covered with a strip of rubber (16) to avoid electrostatic
charging of the plastic dish when it is moved. The use of a
heat-shielded microscope lamp is recommended. With this device large numbers of flies can be inspected over a very long
period without the interruption of work for reanesthetizing
and without danger of killing the animals.
Equipment for carbon dioxide anesthesia may be obtained from
Hans Schütz, Apparatebau, In der Halden, CH-8185 Winkel bei
Bülach, Switzerland. The minimum quantity per order is 20
units. For ordering fly stocks see separate list in section
1.6.

Starting new cultures

When starting new cultures, it is important to be sure that
neither the walls of the bottle nor the surface of the medium
are wet. Although the name Drosophila means "dew-loving",
flies tend to stick to wet places and die.
Labeling: Vials and bottles should be labeled with a marking

pen. Water-insoluble ink is preferred because the accidental
erasing of labels is prevented. On the other hand, the ink
should not create too much of a problem when washing the
glassware. Usually suitable pens can be found in local
stores. If preferred, glued labels can be used so long as
the glue is water-soluble and labels can be soaked off easily.

It is recommended that each basket of cultures be labeled
with a card indicating the cross and the date when the cultures were started before they are put into the incubator or
culture room.

<u>Transfer of flies</u>: The easiest way to start new cultures is
to shake unanesthetized flies directly from one bottle into a
fresh one. First the flies are shaken to the bottom of the
bottle, then both bottles are opened, the bottle with the
flies is put upside down on the empty one, and then the flies
are shaken into the new bottle.
For the beginner a major problem is to use a small enough
number of flies. Too many parental flies yield overcrowded
cultures with small, slow growing larvae and delayed hatching
of F_1 flies.
The direct transfer of flies is an everyday technique but
from time to time the stocks should be checked under the stereomicroscope for the correct phenotype. This should be done
about every third generation.

<u>Pair matings</u>: When anesthetized flies are to be used to
start a new cross, they must be transferred carefully to prevent them sticking to the medium or other damp surface. This
is most easily done by brushing them gently onto the dry wall
of the food vial or bottle while it is held on its side.
After stoppering the container, it should be left in a horizontal position until the flies have revived. Anesthetized
flies can also be put into a clean empty vial to revive and
then shaken carefully into the food bottle.
It is also possible to add flies to a vial containing unanesthetized flies. For this purpose the flies in the vial are
shaken down to the medium, the stopper is pushed to one side
so that a passage is formed through which an anesthetized fly
(held by the wings or legs with a forceps) can be carefully

introduced into the vial. Care has to be taken that the anesthetized flies do not stick to the medium.

Stock keeping

For rearing stocks with large numbers of flies we use bottles with corn meal medium and live yeast. To start cultures one needs well-fed adult flies which produce a lot of fertilized eggs. These are obtained by keeping freshly hatched flies for 2 to 4 days in well-yeasted "feeding" bottles. To stimulate oogenesis the flies need protein-rich food (live yeast). To propagate normally fertile stocks about 10 to 25 pairs are used per 200 ml bottle. Beginners should count the flies; later on with some experience, the number of flies can be estimated. Remember: Avoid overcrowding. Each bottle should be labeled and dated. The parental flies should be removed from the bottles after 2 to 3 days. They may well be used to start a second culture. After removal of the flies from a culture bottle, the medium contains eggs, embryos and young larvae. Depending on the quality of the medium, it may be desirable to add more live yeast. In order to increase the surface and to regulate humidity within the bottle, a small (ca. 10 x 10 cm) piece of absorbent paper (paper toweling) may be added. Finally, the bottles are kept at room temperature or preferably in an incubator at 25°C and 60% relative humidity.

Stock collection

Usually we maintain the stocks during the time of the year when no experiments are conducted. All stocks described in this book can be kept at a low temperature (i.e. 18°C). Since the generation time is longer at a low temperature, stock keeping is less laborious under these conditions. Vigorous stocks can be maintained permanently at 18°C. In this case up to 50 pairs are allowed to lay eggs for at least 4-5 days. This guarantees an optimal population density. After removing the parental flies, we also add yeast and paper. The duration of a generation will exceed 3 weeks. Based on our experience, we recommend keeping three bottles of every stock. Stocks should be inspected for the correct phenotypes about every third generation. Using three bottles which represent separate lines, a contaminated line or a line with lost markers can be replaced by splitting one of the duplicate lines.

Mite control

Mites are probably the most serious contaminants of Drosophila cultures. Check all new stocks for the presence of mites! They should be kept in quarantine for at least two generations before adding them to your other stocks. For more details on mite control measures see Roberts (1986).

Culturing pair matings

For matings with only one or a few females, we use small vials. Care has to be taken that the progeny are limited to a reasonable number. This is usually achieved by limiting the egg-laying period to a maximum of 4 days. Taking out the parental flies after about 4 days has the additional advantage that upon hatching of the progeny no "wrong" flies are present, and that no backcrosses to parental flies are possible which could disturb the successive crosses. The addition of yeast and absorbent paper is recommended for vials just as for bottles.

Collection of virgins

For many experiments, virgin females which have not yet mated are required. In order to collect virgins, select cultures, preferably 9-10 days old, with dark brown pupae ready to hatch and remove all adult flies! Not a single male or female should remain in the culture. Depending on the stock, flies can be collected during the next 6 to 10 hours. At the end of this period, the flies are anesthetized. Under the stereomicroscope, females and males are separated carefully (for the sex characteristics see chapter 2.4) using a fine brush. Usually the males are discarded (or removed to be used in other crosses). Now the females are inspected again to make sure that not a single male has been overlooked. The virgins are then kept in well-yeasted culture bottles without

males. This allows the females to mature sexually, and
oogenesis is stimulated by the protein-rich food. The optimal age for using the virgin females is 3–5 days. Since the
collection of virgins is the most critical part of a successful Drosophila experiment, the individual steps are listed
again below:
(1) Before starting the collection of virgins, not a single
fly must be left behind in the culture.
(2) The two sexes have to be separated very carefully. The
sex combs of the male and the external genitalia are the best
characteristics to use for newly hatched flies.
(3) Before moving the females to culture bottles, make sure
that not a single male is present.
(4) When starting crosses, check a last time for the absence
of males from the group of virgins used. This may be done
very quickly, because at this stage the body colors of the
flies have developed fully. If a male is found, the entire
group of females has to be discarded. The consequences of
such errors can be minimized, if only small groups of females
are matured together in one container.

<u>Collection of males</u>

Males may be collected at any time. The males used should
not be too young. In general we keep the males used for
crosses in sexual isolation for 1 to 2 days and keep them
well fed.

1.6 List of Genetic Symbols and Stocklist

In these lists, a somewhat simplified notation of alleles is
used. Wherever possible, the superscripts used in scientific
papers to identify individual alleles are deleted. In those
cases in which different alleles at one locus lead to dis-
tinctly different phenotypes, superscripts are retained (e.g.
with the white locus).

Symbol	Map position	Description
b	2-48.5	black; black body color
B	1-57.0	Bar; homozygous B/B ♀ and hemizygous B/Y ♂: eyes reduced to narrow bar; heterozygous B/+ ♀: kidney-shaped eyes
B^S	1-57.0	Bar of Stone. Named for a Drosophila geneticist, this allele of B leads to an extreme reduction of the eyes. It is often used to mark the long arm of the Y chromosome in stocks in which it has been translocated to the Y
bw	2-104.5	brown; brownish red eye color
c	2-75.5	curved; wings curved downwards
cn	2-57.5	cinnabar; shiny light red eye color
ct	1-20.0	cut; wings cut to points and edges scalloped
cv	1-13.7	crossveinless; wings lacking cross-veins
Cy	2-6.1	Curly; wings curled upward; homozygous lethal
dm	1-4.0	diminutive; bristles shortened and thin; dm/dm females are sterile
dp	2-13.0	dumpy; posterior part of the wing truncated
e	3-70.7	ebony; black body color
ey	4-2.0	eyeless; eye approx. 1/4 of the normal size
f	1-56.7	forked; bristles shortened, gnarled and bent
flr	3-38.8	flare; wing hairs reduced to amorphous clump of matter; homozygous lethal
H	3-69.5	Hairless; various bristles, especially postverticals and abdominals, missing; homozygous lethal
j	2-48.7	jaunty; curved wings
m	1-36.1	miniature; wings short, about the same length as abdomen
mei-9	1-6.0	meiotic mutant no. 9; disturbance in

		meiosis and in DNA excision repair
mwh	3-0.0	multiple wing hairs; wing cells contain groups of 2-5 hairs
pol	4-3.0	abbreviation for spa^{pol}
p^p	3-48.0	pink-peach; pink-peach eye color
Pm	2-104.5	Plum; brown-dominant; brownish eye color; homozygous lethal
px	2-100.5	plexus; additional veins on the wings
ri	3-47.0	radius incompletus; wing veins do not extend to wing margin
ry	3-52.0	rosy; dark red eye color
Sb	3-58.2	Stubble; reduced bristles; homozygous lethal
sc	1-0.0	scute; missing scutellar bristles
sd	1-51.5	scalloped; abnormal edge of the wing
se	3-26.0	sepia; brown eye color, darkening with age
Ser	3-92.5	Serrate; wings notched at tip; homozygous lethal
sn	1-21.0	singed; bristles short and kinked
spa^{pol}	4-3.0	sparkling-polished; smooth eye surface, eye size slightly reduced
ss^a	3-58.5	spineless-aristapedia; arista replaced by leg-like structure
st	3-44.0	scarlet; shiny light red eye color
sv	4-3.0	shaven; bristles reduced to small stumps
v	1-33.0	vermilion; shiny light red eye color
vg	2-67.0	vestigial; wings reduced
w	1-1.5	white; white eye color
w^a	1-1.5	white-apricot; apricot colored eyes
w^{co}	1-1.5	white-coral; light dull red eye color
X		X chromosome; 1st chromosome, sex chromosome
X^{c2}		X closed no. 2; ring-X chromosome
y	1-0.0	yellow; body color of the adult flies yellow; larval setae and mouth parts light brown
Y		Y chromosome; sex chromosome which normally occurs only in males
+		wild type. When written as a

superscript to any symbol, indicates
the normal (wild type, nonmutant)
allele. This notation is often
abbreviated by using only the +
without the symbol, e.g. w/+ = w/w$^+$.

Multiple mutants with special phenotype

bw ; st p^p	white eye color
cn ; bw	white eye color
cn ; ry	light orange eye color
st p^p	yellow eye color
v ; bw	white eye color
w ; se	white eye color
w^{co} ; se	dark brown eye color
w^{co}/w ; se/se	intermediate between w ; se and w^{co} ; se

Stocklist

The stocks given in the following list may be obtained from:

Mid-America Drosophila Stock Center
Department of Biology
Bowling Green State University
Bowling Green   OH
43403   USA

Drosophila Stock Center
Department of Genetics
University of Umea
S-901 87   Umea
Sweden

Stock no.	Genotype	Used in experiment no.
1	wild type strain	2, 3.2, 3.3, 3.6, 3.7, 3.8, 4.4, 5.1, 5.2, 6.2, 6.3, 6.4, 7.1, 7.2
2	vg	3.1, 3.5, 4.1, 6.2
3	e	3.1, 3.5, 4.7
4	w; ct	3.3

5	w	3.2, 4.4
6	Cy/Pm; H/Sb	3.5, 7.2
7	b	3.5, 4.7
8	se	3.5, 4.4, 4.7
9	spa^{pol}	3.5
10	w m f	3.6
11	C(2L)RM, j; C(2R)RM, px	3.7
12	C(2L)RM, b; C(2R)RM, cn	3.7
13	B	3.8
14	$mei-9^{L1}$	3.8
15	$y\ w^a\ mei-9^a$	3.8, 5.3
16	$y\ mei-9^{L1}\ cv/y^+\ Y\ B^S$	3.8
17	ss^{a40a}	4.2
18	ry^2	4.4
19	w^a	4.4
20	bw	4.3
21	st	4.3
22	$st\ p^p$	4.3
23	v; bw	4.6
24	cn bw	4.6
25	$w^+(P)$	4.8
26	w (M)	4.8
27	Basc	5.1, 7.2
28	$bw;\ st\ p^p$	5.2
29	y w	5.3
30	$X^{c2},\ y\ f/y^+\ Y\ B^S$	5.3
31	mwh	5.4
32	$flr^3/TM3,\ Ser$	5.4
33	w; se	5.4
34	$w^{co}\ sn;\ se$	5.4
35	y	6.4

1.7 General Books on Drosophila

Summaries, reviews and compilations on diverse fields of Drosophila genetics are found in:

ASHBURNER, M.: Drosophila: A Laboratory Manual. Cold Spring Harbor NY: Cold Spring Harbor Laboratory Press 1989.

ASHBURNER, M.: Drosophila: A Laboratory Handbook. Cold
    Spring Harbor NY: Cold Spring Harbor Laboratory Press
    1989.
ASHBURNER, M., NOVITSKI, E. (eds.): The Genetics and Biology
    of Drosophila. Volume 1 a, b, c. New York, London: Aca-
    demic Press 1976.
ASHBURNER, M., WRIGHT, T.R.F. (eds.): The Genetics and Bio-
    logy of Drosophila. Volume 2 a, b, c, d. New York, Lon-
    don: Academic Press 1978, 1979.
ASHBURNER, M., CARSON, H.L., THOMPSON Jr., J.N. (eds.): The
    Genetics and Biology of Drosophila. Volume 3 a, b, c, d,
    e. New York, London: Academic Press 1981, 1982, 1983,
    1986.
HASKELL, G.: Practical Heredity with Drosophila. Edinburgh,
    London: Oliver and Boyd 1961.
LINDSLEY, D.L., GRELL, E.H.: Genetic Variations of Drosophi-
    la melanogaster. Washington: Carnegie Institution of
    Washington, Publ. No. 627, 1968. (New edition in
    preparation by Lindsley, D.L. and Zimm, G.)
RANSOM, R. (ed.): A Handbook of Drosophila Development. Am-
    sterdam, New York, Oxford: Elsevier 1982.
ROBERTS, D.B. (ed.): Drosophila. A Practical Approach. Ox-
    ford, Washington DC: IRL Press 1986.

The yearly newsletter "DROSOPHILA INFORMATION SERVICE (DIS)"
provides fast communication between Drosophila workers. This
is an informal means for exchanging information on new mutants, new and improved techniques, current research inter-
ests, directories, etc. At present, DIS is edited by J.N.
Thompson, jr., Department of Zoology, University of Oklahoma,
730 Van Vleet Oval, Norman, OK 73019-0235, USA. The four
sections of the new text of "Genome of Drosophila melanogaster" by D.L. Lindsley and G. Zimm have been published in
1985, 1986, 1987 and 1990 as numbers 62, 64, 65 and 68 of DIS.

Bibliographies

The first genetic publications concerning Drosophila melanogaster date back to 1910. During the following decades the

experimental work, initially concentrated on Mendelian genetics, was extended to numerous new research areas. Therefore,
the scientific literature contains an enormous base of knowledge on this diverse research field and is available to ev-
erybody. Most of the original literature is in English.
Between 1910 and 1974 some 25 000 papers were published. The
location of older references is facilitated by a number of
bibliographies published as books or special sections in DIS.

MORGAN, T.H., BRIDGES, C.B., STURTEVANT, A.H.: The Genetics
    of Drosophila. Bibliographia Genetica, Vol. II. The
    Hague: Martinus Nijhoff 1925.
MULLER, H.J.: Bibliography on the Genetics of Drosophila.
    Edinburgh: Oliver and Boyd 1939.
HERSKOWITZ, I.H.: Bibliography on the Genetics of Droso-
    phila, Part 2. Slough, Bucks, England: Commonwealth
    Agricult. Bureau, Farnham Royal 1952.
HERSKOWITZ, I.H.: Bibliography on the Genetics of Droso-
    phila, Part 3. Bloomington IN: Indiana Univ. Press 1958.
HERSKOWITZ, I.H.: Bibliography on the Genetics of Droso-
    phila, Part 4. New York: McGraw-Hill 1963.
HERSKOWITZ, I.H.: Bibliography on the Genetics of Droso-
    phila, Part 5, Part 6. New York: MacMillan 1969, 1974.
HERSKOWITZ, I.H.: Drosophila Bibliography, Drosophila In-
    form. Serv. 52, 186-226 (1977).
HERSKOWITZ, I.H.: Drosophila Bibliography, Drosophila In-
    form. Serv. 53, 219-244 (1978).
HERSKOWITZ, I.H.: Drosophila Bibliography, Drosophila In-
    form. Serv. 55, 218-262 (1980).
HERSKOWITZ, I.H.: Bibliography on Drosophila, Drosophila
    Inform. Serv. 56, 197-258 (1981) (contains title index).
HERSKOWITZ, I.H.: Drosophila Bibliography, Drosophila
    Inform. Serv. 58, 227-270 (1982).
HERSKOWITZ, I.H.: Bibliography on Drosophila, Drosophila
    Inform. Serv. 59, 162-257 (1983) (contains title index).
    This volume concludes the bibliography on Drosophila.

2. Morphology of Drosophila Melanogaster

2. Morphology of *Drosophila melanogaster*

The aim of the following studies is to become familiar with
the life cycle and the different developmental stages of Drosophila melanogaster. Every section starts with the presen-
tation of some morphological, developmental and biological
facts. This is followed by practical studies on the morphology and developmental biology of the different developmental
stages.

Material needed: In addition to the usual material (stereomicroscope, etc.) the following are needed for these investigations: a wooden spatula, a few slides, household bleach,
e.g. Chlorox (a 3 % aqueous solution of sodium hypochlorite),
small pipettes with smooth tips, stainless steel dissecting
needles, a beaker with tap water, watch glasses or other
small, shallow containers.

2.1 Eggs and Embryos

External characteristics of the eggs (Figure 5): They are
oval in shape, about 0.5 mm long and 0.2 mm in diameter. The
dorsal side is slightly flatter than the concave ventral
side. At the anterior end of the dorsal side, there are two
filaments. These prevent the eggs from sinking into a wet
medium and provide the vital oxygen supply. The filaments
are extensions of the chorion, the protective egg shell cov-
ering the whole egg. The chorion and the filaments are se-
creted by the follicle cells surrounding the egg during
oogenesis. At higher magnification the honey-combed impres-
sions left by the follicle cells are visible on the surface
of the chorion. Inside the chorion there is a spongy layer
filled with air. It is the reflection of light from this
layer that gives the egg its milky white appearance. Eggs
can be dried by absorbent paper or very carefully with hot
air; care must be taken to avoid desiccation of the interior
of the egg, which would then collapse. Dried eggs can be
made transparent by submersing them in mineral oil. Another
possibility is the digestion of the chorion described below.

36

First let us review how the eggs are formed during oogenesis: Oogenesis takes place in the ovarioles of the ovaries. Ovaries of wild type flies consist of more than 20 ovarioles so that a normal fertile female has over 40 ovarioles, all producing eggs. Oogenesis starts with a diploid oogonium.

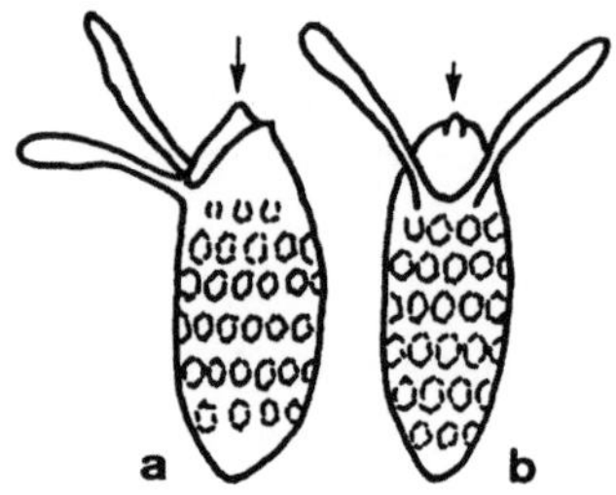

Figure 5. Eggs of
Drosophila melanogaster
a) lateral, b) dorsal.
The arrow points to the
micropyle. After Demerec
and Kaufmann (1965).

This cell divides mitotically to form a cluster of 16 cells. Plasma bridges connect individual cells within the cluster. As the cluster grows, the so-called egg chamber is formed. It consists of one oocyte, 15 nurse cells and a layer of follicle cells. The nurse cells contribute substantially to the growth of the oocyte by supplying yolk. At the end of the growth phase the oocyte reaches about 100 000 times its initial volume. Finally the nurse cells degenerate and the follicle cells secrete the complex chorion. Within a few days the oocyte is mature. Meiosis has started but stops at metaphase I. Inseminated females will soon fertilize the egg and deposit it. Usually, virgin females accumulate one or two mature oocytes in every ovariole. For this reason, well-fed virgins show a whitish swollen abdomen. All these eggs are deposited within a short time after the virgin has mated. This is a major reason for the success of cultures started with well-fed virgin females aged for a few days. If kept without males for extended periods, virgin females will lay unfertilized eggs.
In the testes of the males, haploid sperm are produced from diploid spermatogonia. The process of spermatogenesis goes through two phases: During the first phase, spermatogonia develop into spermatocytes which go through the two meiotic divisions resulting in haploid spermatids. The second phase, spermiogenesis, follows. During this phase the globular early spermatids are transformed into highly specialized mature

sperm with a head containing the highly condensed chromatin and a tail responsible for the mobility of the cell. The differentiation of spermatids into sperm takes place synchronously in a tightly packed group of cells included in a cyst. Every cyst produces a bundle of 64 mature sperm. The mature sperm have an exceptional morphology; the long sperm head containing the genetic material is about 13 μm long and 0.5 μm wide. The tail is more than 100-fold longer than the head, making the sperm a structure 1.7 to 1.8 mm long!

During copulation, the sperm are transferred from the testes into the uterus of the female. From there they swim to special storage organs, the receptaculum seminis and the paired spermathecae. Sperm remain viable in these organs for many days.

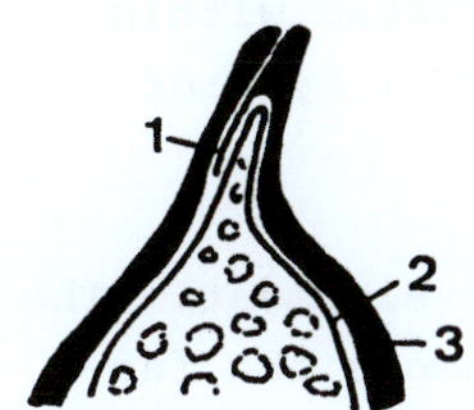

Figure 6. Anterior tip of the egg.
1 = micropyle,
2 = yolk membrane,
3 = chorion.

The eggs are fertilized in the uterus of the females. An individual egg is released from an ovariole into the uterus. Its orientation fits the anterior pole with the micropyle (Figure 6) near the openings of the receptaculum and the spermathecae. During fertilization one single sperm enters the egg through the micropyle. In contrast to earlier views, it has now been shown experimentally that monospermy is the rule in Drosophila. Meiosis of the maternal oocyte nucleus was blocked at metaphase of meiosis I at the end of oogenesis. Fertilization triggers the continuation of meiosis and within about 15 minutes meiosis is completed. These meiotic divisions take place within a cytoplasmic island located on the dorsal side of the egg in the area where the filaments start from the chorion. Maternal meiosis results in the production of one haploid generative nucleus and three polar bodies in the dorsal cortical plasm of the egg. The generative nucleus moves towards the axis of the egg and is trans-

formed into the maternal pronucleus after going through an
interphase-like stage. The polar bodies disintegrate.
At the same time that female meiosis is completed, the initially compact sperm head goes through a phase of striking
morphological changes; the chromatin in the sperm head loos-
ens up and transforms into the interphase structure, which
results in the paternal pronucleus. This transformation
takes place while the paternal nucleus is moving into the
interior of the egg. By the end of the first quarter of an
hour after fertilization, the two pronuclei (one maternal,
one paternal) are located in a cytoplasmic island on the egg
axis. The distance from the micropyle is about 0.2 mm. The
two pronuclei do not fuse at this stage. Side by side, they
synchronously enter the first cleavage division (gonomery).
Each nucleus forms half a mitotic apparatus. It is only at
the end of the first cleavage division, during telophase,
that maternal and paternal chromosomes are enclosed within
one nuclear envelope. This represents an unusual type of
karyogamy.

Figure 7 gives an overview of the embryonic development. The
cleavage divisions in the Drosophila egg are among the fas-
test mitotic divisions known among higher animals: At 25°C
one nuclear cycle takes only about 8 minutes! The cleavage
divisions are syncytial, which means that nuclear divisions
take place without concomitant formation of cell membranes.
Each nucleus surrounded by a cytoplasmic island represents a
so-called energid.
The first 9 synchronous divisions result in some 512 nuclei
initially distributed evenly throughout the egg volume. The
majority of these nuclei migrate to the periphery of the
egg. At the anterior pole about 18 nuclei enter the pole
plasm. These subsequently form the pole cells which will
multiply independently of the rest of the embryo, and some of
which will form the germ cells.
Another group of about 100 nuclei remains in the interior
yolk mass representing the so-called yolk nuclei. These
asynchronously dividing energids will play an important role
in the utilization of the yolk material. The roughly 400
nuclei in the cortical cytoplasm go through 4 additional synchronous divisions and finally form the so-called blastoderm

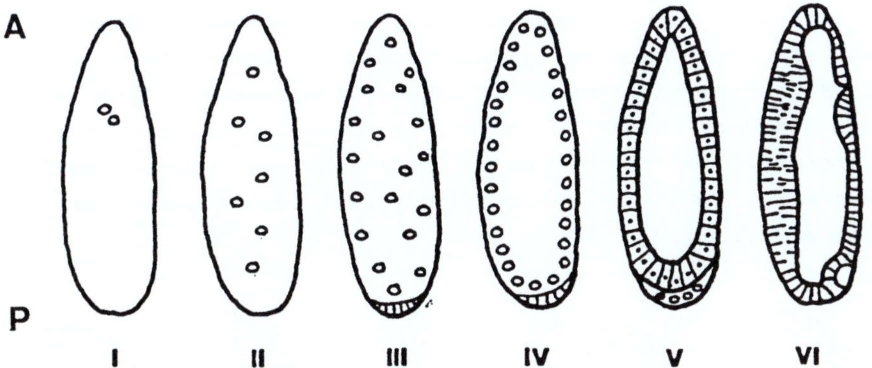

Figure 7. Early embryonic development of <u>Drosophila</u> <u>melano</u>-<u>gaster</u>. After Scriba (1964). A = anterior end, P = posterior end (pole cell region).

I: Fertilized egg after deposition. Pronuclei still un-
      fused.
II: Cleavage. Distribution of nuclei in the egg.
III: Migration of cleavage energids.
IV: Preblastoderm. Nuclei at periphery.
V: Blastoderm. Cell membrane formation.
VI: Beginning of gastrulation.

made up of about 6000 nuclei. When these divisions have ceased the nuclei remain in interphase for about an hour, elongate slightly, and the nucleoli become visible. Finally, cell membranes growing vertically from the surface between the nuclei lead to the formation of the cellular blastoderm. The whole development from insemination to blastoderm forma- tion takes about 3 hours. The pole cells are of special in- terest. They are not included in the blastoderm. They are characterized by a high content of granules, the so-called polar granula already visible in the polar plasm of the ma- ture egg. Some of the cells derived from the polar cells will later form the germ cells in the gonads. The separation of the pole cells from the rest of the cells illustrates the initial step in the separation of the germ line from the soma.

It is interesting to note that the early stages of development are under maternal genetic control. The genetic material of the zygote seems to be largely inactive at first, and
gene products stored in the oocyte before fertilization (and
therefore coded for by the mother's genotype) control early
events. A number of so-called maternal effect genes have
been described.

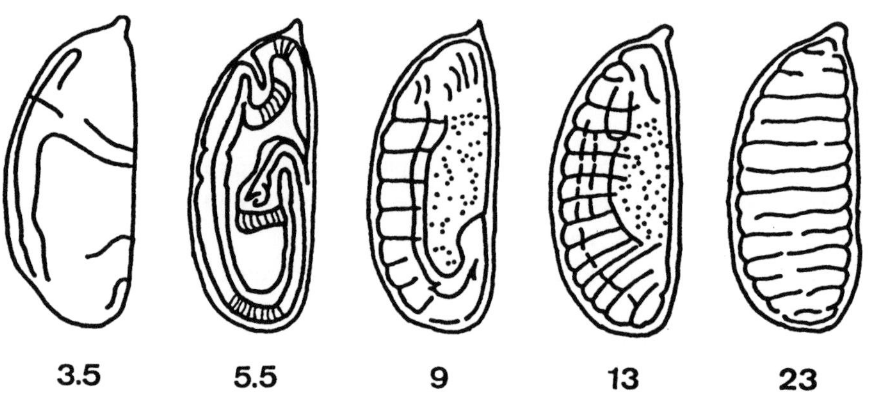

Figure 8. Late embryonic development of Drosophila melanogaster. Simplified after Weber (1974).

   3.5 h embryo : Development of head furrow.
   5.5 h embryo : Invagination and elongation of germ band.
   9   h embryo : Maximal elongation of germ band.
  13   h embryo : Germ band shortening.
  23   h embryo : Segmentation. Larva ready to hatch.

In the blastoderm, about one hour later complex cell movements, cell divisions, invaginations and dislocations of
large cell complexes start (now under the control of the zygote's own genotype). Gastrulation and the following morpho-
genetic events (histogenesis, organogenesis and differentiation) lead to the formation of functional larval organs. An
assortment of certain cell groups (imaginal discs and histoblasts) takes place. These cells will differentiate only
later during the pupal stages. They will form the structures
of the adult fly, the imago.
Recent work with Drosophila using the techniques of molecular
biology is helping to elucidate the genetic control of embryonic and larval development. The use of molecular probes

able to bind to specific gene products has allowed embryologists to localize the activity of a number of genes which function in the organization of the embryo into different body regions, segments and anterior and posterior segment regions. This work with Drosophila is making a large contribution to the solution of very basic problems in regulation of gene control and differentiation.

<u>Preparation of cultures with eggs and Ll larvae</u>: One to two days before the experiment starts, inseminated, actively egg-laying females are transferred to fresh culture bottles. At the start of the experiment, the adult flies are discarded from the bottles. With a wooden spatula or a slightly wet, strong brush, some material is collected from the surface of the medium near the drop of yeast. This material is put into a small drop of water on a slide. Study this material under the stereomicroscope, beginning with a low magnification. Despite the turbidity of the drop due to suspended yeast cells, you should be able to see some eggs (or embryos) and some small Ll larvae (Ll = first larval instar).

<u>Observation of eggs and embryos</u>

In order to make sure that you are seeing the morphological structures of interest make detailed drawings of what you observe and have these checked immediately by your instructor.

<u>Eggs</u>: On the sketch of an egg, you should identify the following details: (1) The anterior pole of the egg with the micropyle, (2) the filaments, (3) the impressions of the follicle cells, and (4) the lateral, ventral, and dorsal side of the egg.

<u>Embryos</u>: To study embryos, it is necessary to remove the chorion. Since mechanical removal is tedious and difficult, we prefer a chemical method. The chorion can be hydrolyzed by submersing the eggs in a 3% sodium hypochlorite solution (Chlorox solution). The easiest way is to brush the eggs into a small volume of bleach solution in a depression in a black glass. Against the black background, the progress of the hydrolysis of the chorion (loss of the white color; for-

mation of small bubbles) can be followed under a stereomicroscope. This usually takes 1 to 2 minutes. To stop the hy-
drolysis, draw the eggs up into a relatively narrow glass
pipette with a smooth tip. The mechanical stress of this
disrupts the remaining layer of the chorion freeing the
eggs. To stop the action of the hypochlorite, the contents
of the pipette is gently squirted into a large volume of tap
water. Now, it is easy to remove the "naked" eggs with a
brush, put them on a slide and study the embryos under a stereomicroscope.

Cleavage stages can be recognized based on the cloudy appearance of the egg content. You should be able to see the blastoderm stages with the superficial layer of nuclei and
cells. The anterior end is characterized by the micropyle
and the posterior end by the pole cells, which are not included in the blastoderm cell layer. In older embryos, you
will see the progressive reduction of the yolk and later on
the appearance of tracheae, partly filled with air, and the
mouth hooks. To identify individual structures within the
embryos, use either the sketches in Figures 7 and 8 or the
figures in various books (Demerec 1965, Ransom 1982, Roberts
1986). Make drawings of different stages showing among other
things: the blastoderm, development of furrows, invagination,
germ band shortening, segmentation, a larva ready to hatch.

2.2 L3 Larvae

The larvae which hatch from the eggs after about one day of
embryonic development are found crawling around the surface
of the medium. They tend to concentrate in the yeast drops
and feed there. During the whole larval devlopment, yeast
cells are the major nutrient. After one day, the first molt
takes place. This is the transition from the first larval
instar (L1) to the second larval instar (L2). After another
day, a second molt leads to the third larval instar (L3).
During L3 the larvae feed for two additional days and grow
tremendously. Older larvae leave the yeast drop and start to
work through the medium. This can be noted by observing the
small tunnels left behind them in the medium. The larvae are

easily visible through the walls of the culture glass with
the naked eye. The first things that one notices are usually
the pigmented mouth hooks which move as the larvae feed.
This makes it easy to check the success of a culture about 3
to 4 days after it was started, just by inspecting the culture vessels from the outside.

Observe the larvae. Externally the body can be seen to be
divided into 12 segments: 1 head segment, 3 thorax segments,
and 8 abdominal segments. The body wall is soft and very
flexible. It consists of an external noncellular cuticle
(with a thin exocuticle and a thick endocuticle) and the cellular epidermis below. At the anterior border of every segment, a ring of small chitinous hooks originating from the
exocuticle can be seen. Their characteristic shape is visible only under higher magnification. A characteristic structure of the head segment are the mouth hooks. With their
dark color they are the most prominent part of the mouth
structures. The chitinous mouth hooks are replaced at every
molt. It is interesting to note that the color of the mouth
hooks is under genetic control; one gene involved is the gene
"yellow" on the X chromosome. The wild type allele (y^+)
determines a dark color, whereas yellow mutations result in
light yellow mouth hooks. This observation is important,
because it demonstrates that certain mutations can already be
detected in larval stages.

The prominent airfilled tracheae extending from the posterior
end along the whole larva to the first thoracic segment can
be identified easily by external examination. The left and
right tracheae branch out into a network of small tubes.
With L3 larvae, the posterior and anterior spiracles functioning as air inlets, are easy to identify. The anterior
spiracles are located on both sides of the first thoracic
segment (prothorax) immediately behind the head segment.
Each spiracle shows 9 characteristic, fingerlike papillae
with terminal openings through which air can enter the tracheal system. The chamber containing the papillae can be
exposed or retracted into the body. The structurally complex
posterior spiracles are yellowish and play the most important
role in gas exchange between the tracheal system and the am-

bient air. This can be demonstrated if larvae are placed in soft, nearly liquid medium. Then the larval body is completely submerged in the medium and only the posterior spiracles extend to the surface.

Additional anatomical details of larvae can be studied in transmitted light. Under the main branches of the tracheae, the white bands of the fat bodies, the coiled gut and the Malpighian tubules are seen. The gonads are located in the 5th abdominal segment (four segments from the posterior end of the larva). The pair of gonads is located within the body cavity near the fat body. The gonads are translucent in contrast to the fat bodies, which appear white. Testes are distinguished from ovaries by size. The ovaries are the size of two to three cells of the fat body, but the testes are much larger. By inspecting some randomly selected larvae, it is easy to learn to differentiate between male larvae with large gonads and female larvae with small, hardly visible gonads. This method of sex determination is important for experiments in which sexed larvae have to be used (e.g. transplantation or cytological preparations).

<u>Preparation of cultures with larvae</u>: 5 to 6 days before the laboratory work is to be done, parental flies are allowed to lay eggs in culture bottles for 2 days. After the flies have been removed, more live yeast is added to the bottles, but no paper. This should provide many well-grown larvae which can be collected for the lab work.

<u>Study of L3 larvae</u>

For the study of the larvae under a dissecting microscope, animals that are at least 4 days old and 4-5 mm long are the most suitable. At this stage it is easy to sex the larvae. In addition, L3 larvae contain tissues suitable for cytological chromosome studies. The brain with the adjacent ganglia is used for preparations of mitotic chromosomes and the salivary glands for analyses of polytene chromosomes.

<u>Dissection of an L3 larva</u>: We shall concentrate on three systems: (1) the salivary glands, (2) the brain and the gan-

glia, and (3) the imaginal discs. Under the dissecting microscope the larva is held near the posterior end with a fine
needle or forceps in a drop of water on a slide. With the
second instrument (needle or forceps) the larva is held just
behind the mouth hooks and then quickly pulled apart. In
this way the larval organs in the head region come free and
can be inspected. The paired salivary glands are connected
to the pharynx and lie alongside the esophagus. They have an
opaque appearance, and the structure of single cells with

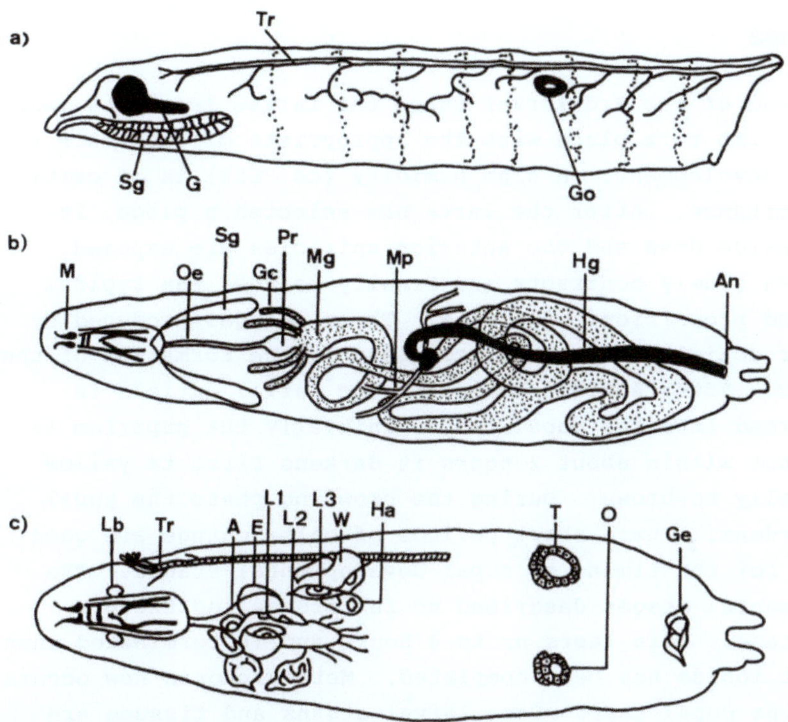

<u>Figure 9</u>. Structure of the L3 larva. a) lateral view, b)
larval organs, c) imaginal discs.
A = antenna; An = anus; E = eye; F = fat body; G = ganglion;
Gc = gastric caeca; Ge = genital plate; Go = gonads; Ha =
haltere; Hg = hindgut; L1, L2, L3 = leg; Lb = labium; M =
mouth hooks; Mg = midgut; Mp = Malpighian tubules; O = ovary;
Oe = oesophagus; Ph = pharynx; Pr = proventriculus; Sg = salivary glands; T = testis; Tr = trachea; W = wing.
After Demerec and Kaufmann (1965) and Demerec (1965).

big nuclei is visible. The brain is situated dorsally near
the anterior end of the salivary glands. It is a paired
structure and easily recognizable by its hemispheric shape.
Below and connected to the brain is the ganglion which extends caudally.
Try to identify the individual imaginal discs according to
Figure 9. In later chapters the origin, structure and development of the eye-antenna and the wing imaginal discs will be
discussed. Make drawings of your observations.

2.3 Pupae

At the end of the 3rd larval stage the larvae leave the medium and climb to a place with the appropriate microclimate for
further development. A high humidity (ca. 65%) is of critical importance. After the larva has selected a place, it
turns upside down and the anterior spiracles are exposed.
The larva slowly contracts and finally reaches the typical
shape and proportions of a pupa. These changes produced by
muscular activity of the cuticle lead to the formation of the
puparium. After an internal molt, the larval cuticle is
transformed into the pupal case. Initially the puparium is
white, but within about 2 hours it darkens first to yellow
and finally to brown. During the browning phase the pupal
case hardens. These short periods of color change are good
markers for the timing of pupal developmental stages. The
developmental stages described so far are called the pre-
pupal stage. This lasts up to 4 hours and is terminated when
the molt inside has been completed. Metamorphosis now occurs
within the pupal case. Some larval organs and tissues are
broken down by histolysis. The salivary glands, the fat
body, the gut, and the muscles are completely decomposed.
Starting from undifferentiated cell groups (the histoblasts)
and the imaginal discs, the structures of the adult body are
built up. The Malpighian tubules and the brain do not disap-
pear but are structurally modified. The legs, wings, eyes,
antennae, and mouth structures as well as the genital apparatus are formed by the differentiation of the corresponding
imaginal discs. The body surface of the head, thorax and

abdomen are formed partly from imaginal discs and partly from histoblasts.

<u>Preparation of cultures with pupae</u>: 8 to 9 days before the practical work, flies are allowed to lay eggs for two days in culture bottles. As with the larval cultures, when the flies are removed, the medium is supplemented with live yeast, but no paper is added.

Study of pupae

Pupae are collected from the wall of the culture bottle with a strong brush. Sometimes, one has to push the pupa gently from the side to be able to remove it from the glass. In order to get different developmental stages, select some light, medium, and dark colored individuals. Under the stereomicroscope, the development can be followed in successive stages. As the pupal development proceeds, more and more imaginal structures become visible. Most prominent are the folded wings, the compound eyes (they change in color from yellow to reddish to dark red-brown), the black segments of the abdomen, the latter being visible only in older pupae. Using needles or forceps, some pupae of different developmental stage may be dissected. Try to identify the different imaginal structures in the different developmental stages. Eyes, legs, and wings are the most prominent structures.

2.4 Adult Flies

The external morphology and anatomy is shown in Figure 11. Before a detailed study of the adult fly, two points of special importance for experimental work will be dealt with. These are: (1) Sex differences between females and males and (2) characteristics of the newly hatched fly, often seen if virgins are to be collected.

(1) Sex differences
For genetic crosses it is mandatory that males and females can be separated from each other without error. Although there are a number of morphological differences between adult

48

females and males, only a few of these can be safely used for
the sexing of flies by beginners.

<u>Sex combs</u>: Put the flies on their backs under the dissecting
microscope and inspect the forelegs. Males have a thick tuft
of hairs on the proximal segment of the tarsus, the so-called
sex comb (see Figure 10). The sex comb is a row of about 10
short, thick, black bristles, located near the distal end of
the front side of the leg. Females lack such a structure.

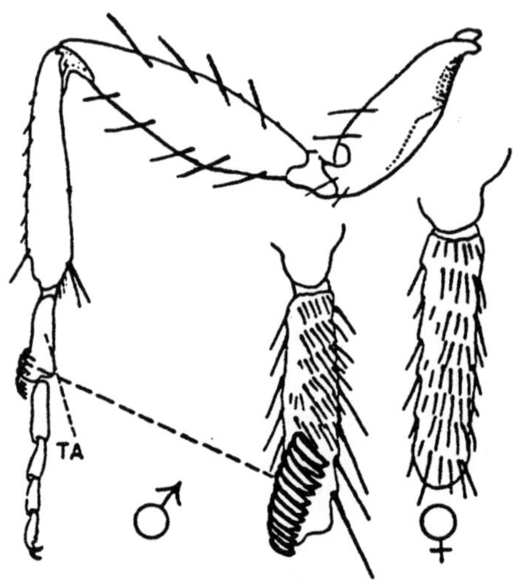

<u>Figure 10</u>. Left foreleg of <u>Drosophila</u> <u>melanogaster</u>. On the
metatarsus (TA) males show the sex comb, which is absent in
females. After Ferris in Demerec (1965).

Once the sex comb has been identified, it can be seen easily
at lower magnification. In our experience, the presence or
absence of the sex comb is the safest characteristic to use
to separate the sexes, especially if newly hatched flies have
to be sexed (see below). A list of additional sex differences is given below:

<u>Genitalia and abdominal coloration</u>: The external genitalia
in the two sexes are very different. The differences can
only be seen at magnification of 25-fold and higher. The
general impression of the male genitalia at lower magnification is a prominent dark structure. Females have 7 abdominal
segments. The dark back edge of every segment is visible in
a dorsal or lateral view. Males have only 5 segments. The
posterior part of the abdomen appears black, because the last

segments are strongly pigmented and form a black region.
These characteristics have to be used with caution for the
identification of the sexes, because newly hatched individuals have not yet developed the characteristic pigmentation
and anesthetized males often extend their abdomens and in
this way the coloration of the abdomen appears striped, similar to that of the females.

<u>Shape of the abdomen</u>: The shape of the abdomen, especially
the posterior end, is clearly different in males and females. In males the posterior end is rounded, whereas in
females the region carrying the anal plates protrudes resulting in a characteristic pointed shape.

<u>Body size</u>: In general, females are larger than males. But
this is not absolute because the actual body size of the
adult flies is strongly dependent on the feeding conditions
during the larval period.

<u>Length of the wings</u>: The wings of the females are longer
than those of the males, but this feature is not suitable for
sexing the adults.

(2) Newly hatched flies

About two hours after the onset of puparium formation a junc-
tion appears near the anterior spiracle which defines a flat
area of the puparium. This "cover" is called the operculum.
When a fly is ready to hatch, its head is positioned below
the operculum. The head has an inflatable saclike structure,
the ptilinum, on the top, and when hatching starts, the fly
opens the operculum by pressing with the extruded ptilinum.
The anterior and lateral junctions open and the fly can crawl
out of the puparium.

The newly hatched fly looks rather strange. It takes about
an hour until the ptilinum retracts and the head takes on its
final shape. Initially, the wings are tightly folded and
appear as a gray, folded mass. Hemolymph is pumped into the
wing veins and the wings slowly unfold. The wings and the
other parts of the body which are soft at first, harden within the next few hours when exposed to the oxygen in the am-
bient air. Newly hatched flies appear larger than older
flies. Especially the abdomen is noticeably extended and may
show a greenish fecal mass (the meconium) which is defecated
soon after hatching. All the colored parts of the fly darken

50

after the fly has hatched. This is first noticeable in the
grayish body color and the black parts of the abdomen. Later, during the first days of the adult life the eye color
also darkens. Certain eye color mutants are more easily
identified in young flies and others in aged flies.
We have discussed the appearance of the newly hatched fly in
detail because they are often seen if virgins are collected.
The beginner has to learn that these individuals are not mutants, but represent a short stage during the normal development of the fly.

<u>Preparation of cultures with adult flies</u>: For the study of
the normal morphology of the adult fly 2- to 6-day-old
individuals from a wild type strain are most suitable. If
cultures are started 9 to 10 days before the lab work, hatching flies will be available. If at the beginning of the lab
work, all flies are removed from the culture bottles, newly
hatched flies will be available within a short time. Hatching normally follows a circadian rhythm with more flies
hatching in the morning than in the evening.

<u>Inspection of adult flies</u>

<u>External morphology of the wild type fly</u>: Examine some well
pigmented individuals (2-6 days old) under a dissecting
microscope at 6x magnification. Sketch details of the body
structures and then identify them by comparison with Figure
11. The major parts of the body are:
<u>Head</u>: Identify the compound eyes, ocelli, antennae with ari-
sta, proboscis.
<u>Thorax</u>: Note the pairs of legs, wings, halteres, scutellum.
<u>Abdomen</u>: Examine the segments (dorsal tergites with a black
posterior edge, ventral sternites), exterior genitalia of
males and females.

<u>Collection of virgin females</u>: Examine newly hatched flies
(less than 6 hours old). Try to see all the various characteristics of the newly hatched flies and the sex differences
as explained above. Separate the females from the males.

2.5 Some Mutants

<u>Material</u>: 2- to 6-day-old flies from various strains are needed. A mixture of flies is prepared from strains which will be used for the different experiments and which carry easily visible mutant traits.

<u>Analysis of mutants</u>

Anesthetize the flies and try to identify the various mutants. Always compare with wild type flies. To facilitate the analysis, a short description of the most important traits is given below.

<u>Body color</u>: The wild type is gray to brown. Mutants can be darker, even black. In other mutants a lighter, yellow body color can be found.
<u>Eye color</u>: Wild type eyes are brick red. Mutants can be brownish or bright red or lighter, i.e. yellowish to white.
<u>Eye shape</u>: In addition to eye color the eye shape and size can also vary. Some examples are given in Figure 12.
<u>Wing size and shape</u>: Among the most easily visible mutants are those that affect wing size and shape. Some examples are shown in Figure 13.
<u>Wing veins</u>: In normal flies the wing has a characteristic pattern of veins. Figure 14 gives two examples of mutations that affect the wing veination.
<u>Bristles</u>: There is a whole series of mutations that influence the shape, size and positioning of the bristles. Selected examples are given in Figure 15.

Make a list and description of the mutants found. Try to find the proper names of the mutants in the list given in chapter 1.6.

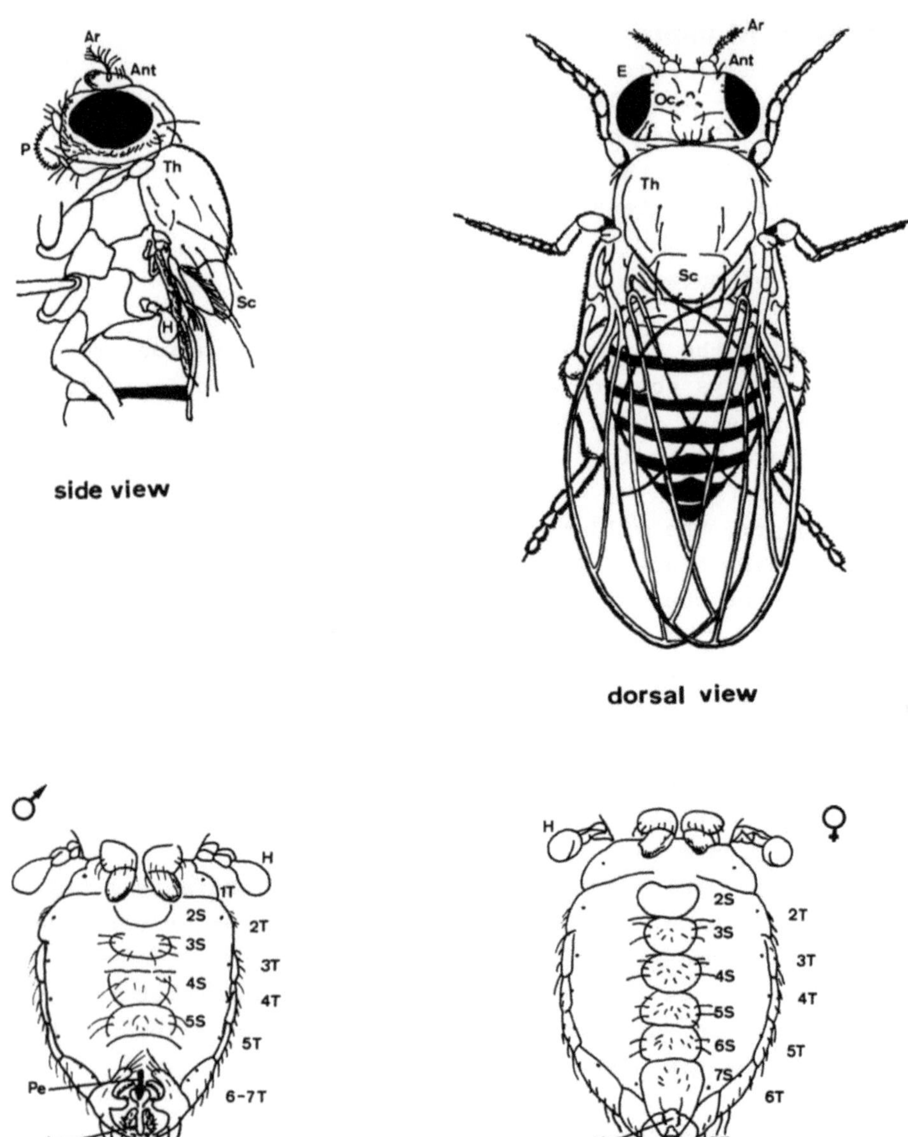

Figure 11. External morphology of adult flies. After Lindsley and Grell (1968). An = anus, Ant = antenna, Ar = arista, E = compound eye, H = haltere, Oc = ocelli, P = proboscis, Pe = penis, S = sternite, Sc = scutellum, T = tergite, Th = thorax, V = vagina.

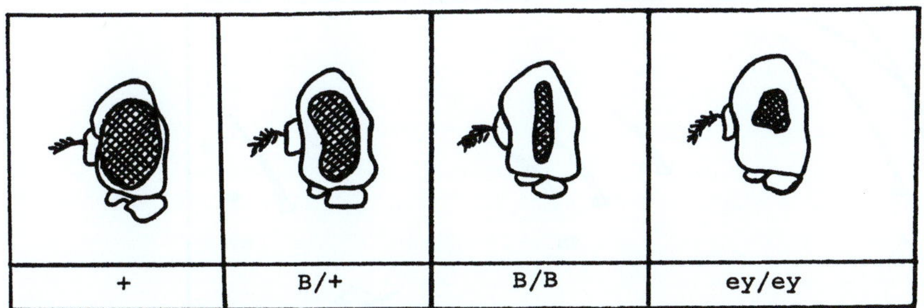

<u>Figure 12</u>. Different eye shapes. + : wild type; B/+ :
Bar heterozygous with kidney-shaped eyes; B/B : Bar homozygous with bar eyes; ey/ey : eyeless homozygous with rudi-
mentary eyes.

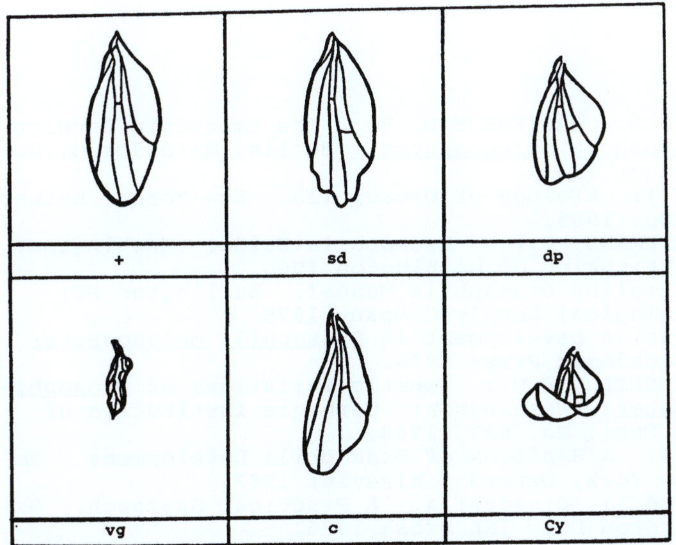

<u>Figure 13</u>. Various wing shapes. After Flagg (1975).

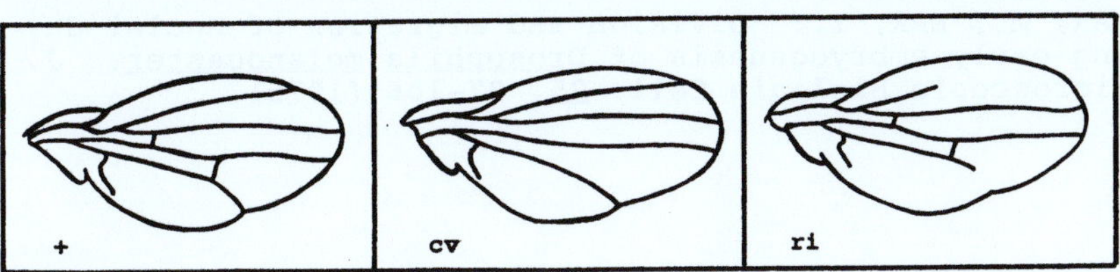

<u>Figure 14</u>. Wing vein mutants. After Flagg (1975).

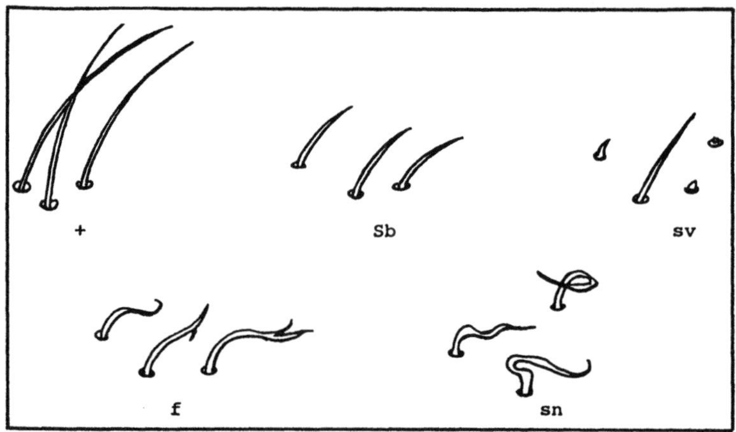

Figure 15. Bristle mutants. After Flagg (1975).

Literature

CAMPOS-ORTEGA, J.A., HARTENSTEIN, V.: The Embryonic Develop-
    ment of Drosophila melanogaster. Berlin, Heidelberg, New
    York: Springer 1985.
DEMEREC, M. (ed.): Biology of Drosophila. New York: Hafner
    Publishing Co. 1965.
DEMEREC, M., KAUFMANN, B.P.: Drosophila Guide. Washington:
    Carnegie Institution of Washington 1965.
FLAGG, R.O.: Carolina Drosophila Manual. Burlington NC:
    Carolina Biological Supply Company 1975.
KING, R.C.: Ovarian Development in Drosophila melanogaster.
    New York: Academic Press 1970.
LINDSLEY, D.L., GRELL, E.H.: Genetic Variations of Drosophi-
    la melanogaster. Washington: Carnegie Institution of
    Washington, Publ. No. 627, 1968.
RANSOM, R. (ed.): A Handbook of Drosophila Development. Am-
    sterdam, New York, Oxford: Elsevier 1982.
ROBERTS, D.B. (ed.): Drosophila. A Practical Approach. Ox-
    ford, Washington DC: IRL Press 1986.
SCRIBA, M.E.L.: Beeinflussung der frühen Embryonalentwick-
    lung von Drosophila melanogaster durch Chromosomenaber-
    rationen. Zool. Jb. Anat. 81, 435-490 (1964).
WEBER, H.: Grundriss der Insektenkunde. Stuttgart: Gustav
    Fischer 1974.
ZALOKAR, M., ERK, I.: Division and migration of nuclei dur-
    ing early embryogenesis of Drosophila melanogaster. J.
    Microscopie Biologie Cell. 25, 97-106 (1976).

3. Transmission Genetics

3.1 Dihybrid Cross with Independent Assortment

Inheritance is the transmission of genetic information from parents to offspring. Most of the genetic information is located on the chromosomes. The principles of chromosomal inheritance were first discovered experimentally by Gregor Mendel in 1865 and rediscovered by Correns, von Tschermak and de Vries in 1900. Mendel's laws describe the basic rules of simple inheritance:

<u>Law of dominance</u>: When purebred (homozygous) strains differing in a particular trait and the corresponding allele pair (a^+a^+ and aa, respectively) are crossed, the F_1 individuals (a^+a) are uniform, regardless of the direction of the cross. The F_1 from reciprocal crosses, either a^+a^+ females x aa males or aa females x a^+a^+ males, usually express only one of the two characteristics (that controlled by the dominant allele a^+) to the exclusion of the other (the recessive one). This is the law of dominance also known as the principle of the uniformity of the F_1.

<u>Law of segregation</u>: Recessive characteristics, which are masked in the heterozygous (a^+a) F_1 arising from a cross between purebred strains, reappear in a specific proportion of the F_2. That is, the members of an allele pair (a^+a) separate from each other without influencing each other, when an individual forms haploid germ cells. This is the principle of segregation.

<u>Law of independent assortment</u>: Members of different allele pairs (e.g. a^+a and b^+b) assort independently of each other when haploid germ cells are formed provided the genes in question are unlinked (located on different chromosomes). This is the principle of independent assortment.

<u>Problem</u>

Flies from two stocks each differ from the wild type in one characteristic (wing size or body color). Test experimentally whether or not the two characteristics result from two

unlinked Mendelian genes. Do this by crossing flies of the
two stocks to obtain an F_1 progeny. Then intercross the
F_1 flies to obtain an F_2 progeny. This will allow the
detection of independent assortment if present.

<u>Material</u>

From one stock, in which all flies have vestigial wings, you
isolate virgin females. All these females are homozygous for
the recessive allele vestigial (vg/vg) which is located on
chromosome 2 at position 67.0 (2-67.0). From the second
stock, in which all flies have a dark, nearly black body col-
or, males are collected. They are homozygous for the recessive allele ebony (e/e). The ebony gene is located on chromosome 3 at position 70.7 (3-70.7).

<u>Experimental work and interpretation of data</u>

Start work by filling in the available information in the
spaces provided below. Remember that the haploid chromosome
set of Drosophila is made up of 4 chromosomes. We assume
that all genes other than those specifically mentioned are
wild type. Chromosomes carrying no gene of particular interest within the context of the present experiment are also
designated "+". As a symbol for the Y chromosome, which in
normal individuals is present only in males, we use the symbol "Y". For the X chromosome we write "X". Now write down
the genetic formulas of the parental flies (P = parental generation). Starting from these formulas, determine the types
of gametes formed by the parents.

<u>Genotypes of the P generation</u>   <u>Phenotype</u>
Chromosome: 1   2   3   4
    ♀:   X/X ; vg/vg ; e^+/e^+ ; +/+   ....
    ♂:   X/Y ; vg^+/vg^+ ; e/e ;   +/+   ....

The P individuals will produce the following kinds of gametes:

♀: ....

♂: ....

Predict the genotypes of the F$_1$ individuals

This is usually done by means of genetic checkerboards (Punnett squares). This is a convenient way of working out all the possible combinations of different types of eggs from the female parent and sperm from the male parent. The steps followed are:

(1) Work out the genotypes of the eggs produced by the female and write each different genotype in one space horizontally along the top of the diagram.

(2) Work out the genotypes of the sperm produced by the male and write each different genotype in one space vertically along the left side of the diagram.

(3) Fill in each square with the genotype of the type of zygote formed by the combination of the egg at the top of the relevant column and the sperm at the left of the relevant row. Note that by convention, in a heterozygote the dominant allele is written first (vg$^+$/vg) no matter which parent donated which allele. Another thing to remember is that each different type of gamete is written only once. The genotype of an individual fly or type of fly is written as if it consisted of only one cell, the zygote (or fertilized egg) from which the individual developed.

Determine the genotype in the different boxes of the checkerboard and give the phenotype and sex of the different classes of progeny to the right of the diagram. We expect the following F$_1$ (first filial generation) progeny:

By interbreeding the F_1 (brother x sister cross) we obtain
the F_2 (second filial generation). In our future work we
will concentrate on the genes which are of particular interest for a given cross. In the present case, these are the
genes determining wing size and body color of the flies. The
simplified notation of the genetic formulas for the cross
leading to the F_2 is:

.... x ....

Working out the different kinds of gametes each parent will form

To predict the results of any genetic cross, the first step
is to work out the genotypes of all the different kinds of
gametes which each parent can form. It is important to remember that each gamete must include one allele of each different gene. Each different combination should be included
only once, and all combinations must be present. A simple
method to make sure of forming all combinations is to make a
branch diagram. A locus which is homozygous produces only
one kind of allele, while one which is heterozygous produces
two kinds. For example the hypothetical genotype a^+/a;
b/b; c^+/c would be diagramed as follows:

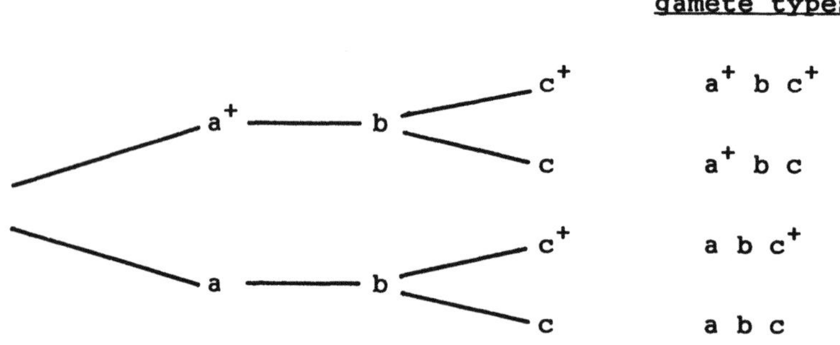

To check that all possible combinations have been included,
apply the formula: no. of gamete types = 2^n (where n = no.
of heterozygous loci). In the example above, three loci are
included but only two are heterozygous, and therefore n = 2
and the number of kinds of gametes is 4.

Now diagram the different kinds of gametes expected from the F_1 individuals in your experimental cross as given above:

The different kinds of gametes are entered into the next checkerboard, which is used to determine the expected genotypes of the F_2 progeny.

♂ \ ♀				

Procedure

The times given below in parentheses refer to the course of the experiment if practical work can only be done weekly.

<u>Day 1 (week 1)</u>: 4-day-old virgins of the vg stock are mated to e males. Prepare two vials with two pairs of flies in each one.

<u>Day 4 (week 2)</u>: The parental flies are removed from the vials.

<u>Day 12 (week 3)</u>: Check the F_1 progeny. Are all the flies phenotypically uniform? The flies do not have to be counted. For the next cross we select at random two pairs of flies and put them into a new vial.

<u>Questions</u>:

(a) Why is it mandatory to remove the P flies from the vials on day 4 or at least before the F_1 hatch?

(b) Is it appropriate to start an F_2 cross with nonvirgin F_1 females? What problems might result if F_1 females have already mated with their F_1 brothers?

<u>Day 15 (week 4)</u>: The parental individuals are removed from the vials.

<u>Day 23 (week 5)</u>: The F_2 progeny are classified according to phenotype and counted.

<u>Questions</u>:

(c) How many different genotypes are expected?

(d) What ratio do you expect for the different phenotypes? Record your answers in the following table.

Genotype					
Phenotype	Wing size				
	Body color				
Expected ratio					
Fly count	Total:				
Observed ratio					

In the experimental work, the flies of the F_2 generation
are classified according to their phenotype and counted. For
each fly the size of the wings and the body color are recorded. The sex of the flies is not registered. The numbers
found are entered in the table.

<u>Questions</u>:
(e) Do you find all the different F_2 phenotypes which are
theoretically expected?
(f) Does the observed ratio of phenotypes coincide with the
theoretical expectation?

To answer the last question, we must analyze our data statistically. The agreement between the expected and the observed
frequency of individuals in the different phenotypic classes
can be measured by a Chi-square test. The Chi-square is defined as:

$$Chi^2 = S [(o - e)^2 / e]$$

The symbols denote:
  S the sum over all classes,
  o the observed number of individuals in a particular class,
  e the expected number of individuals in this class. The
     expectation is based on the prediction found by applying
     Mendel's laws. The expected number is calculated by mul-
     tiplying the expected fraction by the total number of
     flies counted.

In our case with four different classes the Chi-square values
for different probabilities, P, describing the agreement between experiment and expectation are given in the following
probability table:

P	=	0.99	0.975	0.95	0.90	0.80	0.70
Chi^2	=	0.115	0.216	0.32	0.584	1.005	1.424

P	=	0.5	0.3	0.2	0.1	0.05	0.01
Chi^2	=	2.366	3.665	4.642	6.251	7.815	11.34

To have large enough numbers for the statistical analysis the
data from all students are pooled. Using the table shown
below, we calculate the Chi-square value. The Chi-square

value is located in the probability table above and the corresponding P value read. The P value is the cumulative probability of a deviation from the expected as large or larger than that observed. Conventionally, in biology one assumes that P values below 0.05 (= 5%) indicate a significant deviation of the observation from the expectation. The difference is larger than what would be expected to occur in experimental work, simply by chance. In such a case one assumes that the observation is <u>different</u> from the expectation. For further details see textbooks.

Observed numbers (= o) Total = ...				
Expected numbers (= e)				
(o - e)				
(o - e)^2				
(o - e)^2 / e				

$$\text{Chi}^2 \quad = \text{S} [(o - e)^2 / e] = \ldots\ldots\ldots\ldots\ldots$$

$$\text{P} \quad = \ldots\ldots\ldots\ldots\ldots$$

Interpret the observed Chi^2. In case the observation deviates significantly from the expectation, try to find reasons for the deviation. Summarize the result of the experiment in a few sentences.

<u>Literature</u>

MORGAN, T.H.: Random segregation versus coupling in
    Mendelian inheritance. Science <u>34</u>, 384 (1911). In:
    Great Experiments in Biology (eds. M.L. GABRIEL and S.
    FOGEL). Englewood Cliffs NJ: Prentice-Hall 1955.

MORGAN, T.H.: Factors and unit characters in Mendelian
    heredity. American Naturalist <u>47</u>, 5-16 (1913).
STURTEVANT, A.H.: A History of Genetics. New York: Harper
    and Row 1965.

English translation of Mendel's classical work:
PETERS, J.A. (ed.): Classic Papers in Genetics. Englewood
    Cliffs NJ: Prentice-Hall 1959.

Textbooks

BURNS, G.W.: The Science of Genetics, 5th ed. New York:
    Macmillan Publishing Co. 1983.
GARDNER, E.J.: Principles of Genetics, 5th ed. New York:
    John Wiley and Sons 1975.
GOODENOUGH, U.: Genetics, 2nd ed. New York: Holt, Rinehart
    and Winston 1978.
HARTL, D.L., FREIFELDER, D., SNYDER, L.A.: Basic Genetics.
    Boston MA: Jones and Bartlett 1988.
KLUG, W.S., CUMMINGS, M.R.: Concepts of Genetics, 2nd ed.
    Columbus OH: Merill Publishing Co. 1986.
RUSSELL, P.J.: Lecture Notes on Genetics. Oxford, Boston:
    Blackwell Scientific Publications 1980.
STRICKBERGER, M.W.: Genetics, 2nd ed. New York: Macmillan
    Publishing Co. 1976.
WARDLAW, A.C.: Practical Statistics for Experimental Biolo-
    gists. Chichester UK: John Wiley and Sons 1985.
ZUBAY, G.: Genetics. Menlo Park CA: Benjamin/Cummings
    Publishing Co. 1987.

3.2 Sex-Linked Inheritance

Mendel's law of dominance predicts that hybrids from reciprocal crosses are identical and uniform. Due to the hemizygosity of sex-linked genes in males, these genes behave differently in reciprocal crosses.

<u>Problem</u>

Determine the difference in the progeny of reciprocal crosses when a sex-linked (X-linked) gene is analyzed.

<u>Material</u>

A wild type strain with normal red eyes and a strain with white eyes due to the mutation white (w, 1-1.5; genotype of the females X, w/X, w, genotype of the males X, w/Y). Collect virgin females and males from both strains.

<u>Experimental work and evaluation</u>

Carry out the two reciprocal crosses between flies with normal red eyes and flies with white eyes.

<u>Cross A</u>　　　　　　　　　　<u>Cross B</u>

Note: ─────7 is used to represent the Y chromosome.
What types of progeny are expected in the F_1 in the two cases?

<u>Cross A</u>　　　　　　　　　　<u>Cross B</u>

<u>Phenotype</u> (sex and eye color):

       <u>Cross A</u>   <u>Cross B</u>

1   ....

2   ....

Examine the progeny of the two crosses A and B.
Interpret the results:

<u>Literature</u>

BRIDGES, C.B.: Sex in relation to chromosomes and genes.
    American Naturalist <u>59</u>, 127-137 (1925). In: Classic
    Papers in Genetics (ed. J.A. Peters). Englewood Cliffs
    NJ: Prentice-Hall 1959.
MORGAN, T.H.: Sex limited inheritance in Drosophila.
    Science <u>32</u>, 120-122 (1910). In: Papers on Genetics. A
    Book of Readings (ed. L. Levine). St. Louis MO: Mosby
    Comp. 1971.

3.3 Dihybrid Test Cross with Linked Genes

Two genes which are located on the same chromosome are referred to as linked, because they are often transferred together to the progeny. The closer to each other the two
genes are, the closer the linkage is. As a relative measure
for the distance between two linked genes one uses the frequency of meiotic recombination due to crossing over between
the genes. This can be measured in females, which are double
heterozygotes for these genes.

<u>Problem</u>

Determine the map distance between two recessive genes on the
X chromosome.

<u>Material</u>

Collect virgin females from a strain in which all the flies
have white eyes and cut wings. Collect males from a wild
type strain.

<u>Experimental work and evaluation</u>

The cross to be made reads as follows (X and Y are the sex
chromosomes):

<u>Genotypes of the P generation</u>   <u>Phenotype</u>

♀ :   X, w ct / X, w ct   ....

♂ :   X, w^+ ct$^+$ / Y   ....

The P individuals produce the following kinds of gametes:

♀ : ....

♂ : ....

<u>Determination of the F_1 genotypes</u>

<table>
<tr><td>♂ \ ♀</td><td></td></tr>
<tr><td></td><td></td></tr>
<tr><td></td><td></td></tr>
</table>

Phenotype
(eye color, wing form
and sex)

. . . .

. . . .

<u>Test or back cross</u>

We cross the heterozygous F_1 females again with double mutant males, i.e. with males carrying the same markers as the
P females. Since the genes used in this cross are sex-
linked, this can be done by intercrossing the F_1 flies.
The cross reads as follows:

. . . . x . . . .

In <u>Drosophila melanogaster</u> meiotic crossing over between homologous chromosomes takes place only in females. Crossing
over is completely absent in meiosis in the male. Enter the
expected gamete types directly in the following diagram.

<u>Genotypes of the test cross progeny</u>

	Without recombination		With recombination	
♂ \ ♀				

Enter the expected phenotypes in the table below. Classify and count the flies:

Genotype					
Phenotype	Eye color				
	Wing size				
Expected ratio					
Fly count Total:					

Based on the number of flies which show recombination between w and ct, the distance between the two genes can be calculated as follows:

$$\frac{\text{Number of flies with recombination}}{\text{Total number of flies counted}} \cdot (100) = \ldots\ldots\ldots\ldots$$

Interpretation:

Literature

MORGAN, T.H.: Complete linkage in the second chromosome of the male of Drosophila. Science 36, 719-720 (1912).
MORGAN, T.H., STURTEVANT, A.H., BRIDGES, C.B.: The evidence for the linear order of the genes. Proc. Nat. Acad. Sci. (Wash.) 6, 162-164 (1920).

3.4 Sex Determination

Various mechanisms of genotypic sex determination are known.
We shall only deal with the situation found in <u>Drosophila</u>
<u>melanogaster</u>, which is different from that in mammals even
though both have XX females and XY males. As you have already seen in the experiment "Dihybrid cross with independent
assortment", the normal sex chromosome configuration in Drosophila is two X chromosomes in the female and an X and a Y
chromosome in the male. The Y chromosome is completely different from the X chromosome both morphologically and with
respect to its genetic content. Due to the fact that XX females produce only one type of gamete with respect to the sex
chromosomes (X-bearing oocytes), they are called homogametic. The XY males are heterogametic because they produce two
different types of gametes (X-bearing and Y-bearing sperm).
Bridges (1925) showed that the presence or absence of the Y
chromosome does not determine the sex. Because female-deter-
mining genes are located on the X chromosome and male-deter-
mining genes on the autosomes, sex in Drosophila is determined by the ratio between the number of X chromosomes and
the number of sets of autosomes. This ratio can be expressed
by the sex index, I, which is defined as follows:

$$I = \frac{\text{Number of X chromosomes}}{\text{Number of sets of autosomes}}$$

A set of autosomes consists of the haploid set of autosomes
(one each of chromosomes 2, 3 and 4). Based on the sex index, I, the sex is as follows:

I	Sex
> 1.0	meta female
1.0	female
1 > I > 0.5	intersex
0.5	male
< 0.5	meta male

Unlike the situation in higher animals, where secondary sexual characteristics are usually determined by levels of cir-
culating sex hormones, in Drosophila the sex of each cell is

determined by its chromosome constitution. It is possible to
get exceptional individuals where a sector or even a half of
the body is one sex and the remainder the other sex (gynan-
dromorphs). For example an XX individual which had lost one
X chromosome in a cell in the early embryo, would have a
clone of cells in the body in which the sex index was 0.5
(male) in an otherwise XX female body (I = 1).

To illustrate this rule we can diagram three different crosses. This will show further phenomena in connection with
chromosome segregation. In these three crosses we will only
consider the segregation of the sex chromosomes and assume
that the autosomes are diploid.

a. Compound chromosomes

Drosophila strains are available in which either two X chromosomes or an X and a Y chromosome are stably joined in a
so-called compound chromosome. These compound chromosomes
are also called attached-X ($\overline{XX}$) and attached-XY ($\overline{XY}$), respectively. As these compounds are stable, two new types of ga-
metes are produced in meiosis as compared to the normal segregation of two free sex chromosomes: gametes which contain
the compound chromosome and gametes lacking any sex chromosome.
The following cross is performed:

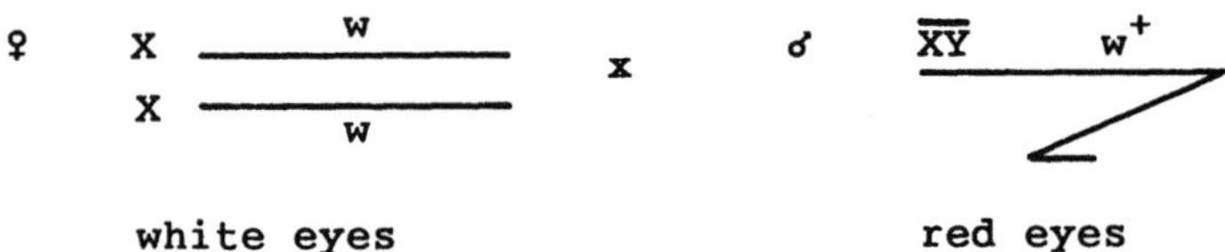

The X chromosomes of the female carry the allele white (w,
1-1.5), the attached-XY of the male contains the normal wild
type allele w^+. Remember that a normal Y chromosome does
not carry an allele of w^+.

Gametes of the female: ....

Gametes of the male: ....

Work out the F_1 genotypes:

<table>
<tr><td>♂ \ ♀</td><td></td></tr>
<tr><td></td><td>1</td></tr>
<tr><td></td><td>2</td></tr>
</table>

Predict the eye color, and with the help of the sex index, I, the sex of the progeny.

	I	Sex	Eye color
Phenotype: 1	....		
2	....		

The progeny obtained are red-eyed females and white-eyed males. The females contain an extra Y chromosome in addition to the two X chromosomes ($X\overline{XY}$); the males, however, contain only one X chromosome (XO). Note that the absence of a Y chromosome is denoted by an "0". Females with an additional Y chromosome are viable and fertile, whereas XO males are viable but sterile.

b. Primary nondisjunction

Primary nondisjunction is the rare failure of segregation of two homologous chromosomes in meiosis. The spontaneous nondisjunction frequency is approximately 5×10^{-5} per chromosome in Drosophila. Such a nondisjunction event leads to the formation of one gamete with both homologous chromosomes and of another without the chromosome concerned. We can illustrate the consequences of primary nondisjunction in female meiosis with the help of a second hypothetical cross:

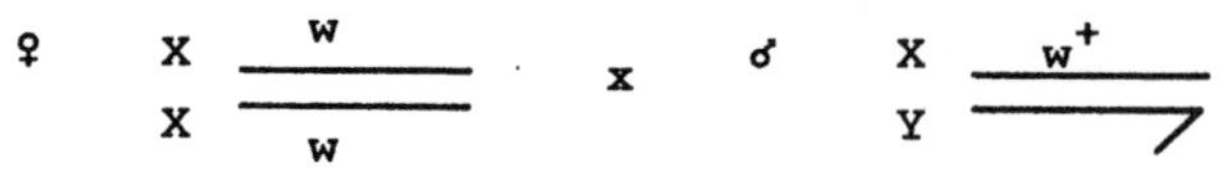

$$\text{♀} \quad X \ \frac{w}{\underline{\quad\quad}} \quad \times \quad \text{♂} \quad X \ \frac{w^+}{} \\ X \ \overline{\underline{\quad w \quad}} \qquad\qquad\qquad Y$$

white eyes   red eyes

Gametes of the female: normal: ....

                         nondisjunction: ....

Gametes of the male: normal: ....

	normal	nondisjunction	
♂ \ ♀			
	1	3	5
	2	4	6

		I	Sex	Eye color
Phenotype:	1	....		
	2	....		
	3	....		
	4	....		
	5	....		
	6	....		

In addition to the normal red-eyed XX females and white-eyed
XY males, the following flies will occur rarely as a consequence of primary nondisjunction: Red-eyed X0 males and
white-eyed XXY females. The combination Y0 is lethal; in
Drosophila all sex chromosome configurations without an X
chromosome are lethal. The combination XXX leads to a meta
female; these are, however, practically lethal (survival rate
less than 0.5%). A surviving meta female would be sterile.

c. Secondary nondisjunction

This term describes the normal segregation of chromosomes in
a trisomic (2n + 1) genotype where an extra chromosome is
present. In meiosis of such an individual four different
types of gametes are produced. As an example we will analyze
the cross between a white-eyed XXY female from the second
cross and a normal male.

$$\female \quad \begin{array}{l} X \;\underline{\quad w \quad} \\ X \;\underline{\quad w \quad} \\ Y \;\underline{\qquad} \end{array} \qquad x \qquad \male \quad \begin{array}{l} X \;\underline{\quad w^+ \quad} \\ Y \;\underline{\qquad} \end{array}$$

white eyes   red eyes

Gametes of the female: ....
Gametes of the male: ....

$\male$ \ $\female$				
	1	3	5	7
	2	4	6	8

I	Sex	Eye color		I	Sex	Eye color
1	....			5	....	
2	....			6	....	
3	....			7	....	
4	....			8	....	

Eight different sex chromosome configurations are produced.
In this cross two of them are lethal (XXX and YY). The other
six are red-eyed females (XX and XXY), red-eyed males (XY),
white-eyed females (XXY) and white-eyed males (XY and XYY).
The flies which carry an additional Y chromosome are not distinguishable from the other flies.

This secondary nondisjunction plays an important role in the chromosome segregation in a triploid strain.

<u>Problem</u>

<u>Triploid strain</u>: In <u>Drosophila</u> <u>melanogaster</u> triploid females are viable and fertile. A triploid strain is the result of a continuous crossing of triploid females with normal diploid males. In the example below the triploid females contain a compound-X chromosome ($\overline{XX}$) and a free X chromosome in addition to three sets of autosomes. The continuous culture of a triploid strain is difficult and tedious. Therefore, we will work out a problem to illustrate the principles involved. Analyze the following cross; the 4th chromosome is deleted for simplicity.

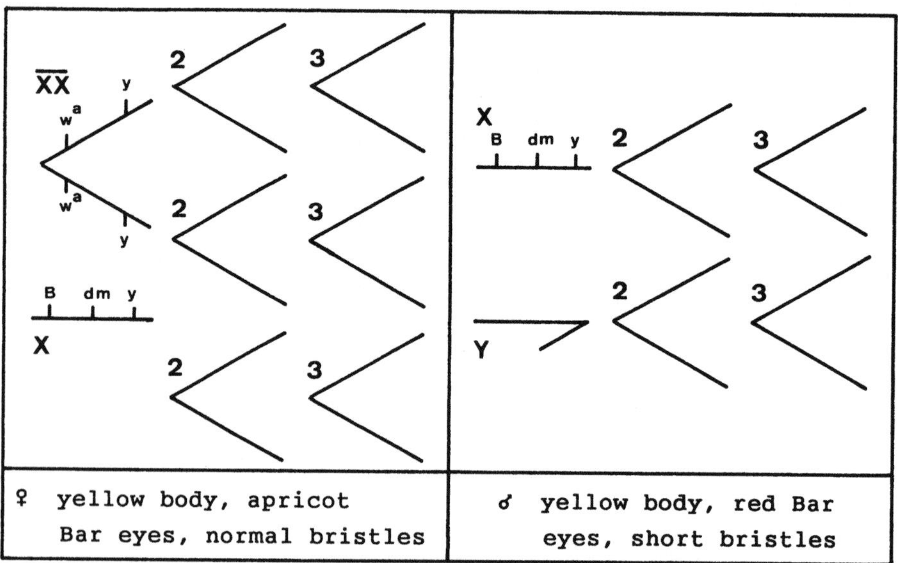

The X chromosomes carry the following markers:

y	yellow (1-0.0); yellow body color.
w^a	white-apricot (1-1.5); apricot eye color.
dm	diminutive (1-4.6); short and thin bristles (dm/dm females are sterile).
B	Bar (1-57.0); Bar eyes.

Depending on the genotype the following eye shapes are
produced:

.+/+	B/+/+	B/+	B/B and B/Y

In this cross the segregation of the large autosomes has to
be considered as well (secondary nondisjunction in triploid
females!). How many types of gametes are produced by the
triploid female? Enter the gamete types immediately in the
checkerboard on the next page and determine the genotypes of
the progeny. Designate all the lethal combinations; take
into account that monosomy for one of the two large autosomes
(2 or 3) is lethal. Such incomplete combinations are called
aneuploid. How many combinations are lethal?

Afterwards determine the expected sex of the surviving
progeny with the help of the sex index, I. Enter the
genotypes in the table below and determine the phenotypes.
Which flies are sterile? Can all the fertile flies be used
for a continuous culture of a triploid strain?

Checkerboard for triploid strain

♀ ♂		

♀   ⟨ ⟨ ⟨ / — ⟨ ⟨ / ⟨ ⟨   ×   — ⟨ ⟨ / → ⟨ ⟨ / ⟨ ⟨   ♂

Genotypes	I	Phenotypes				
		Sex	Eye		Body	Ferti-
			color	shape	color	lity

Literature

BAKER, B.S., RIDGE, K.A.: Sex and the single cell. I. On
    the action of major loci affecting sex determination in
    Drosophila melanogaster. Genetics 94, 383-423 (1980).
BRIDGES, C.B.: Sex in relation to chromosomes and genes.
    American Naturalist 59, 127-137 (1925). In: Classic
    Papers in Genetics (ed. J.A. Peters). Englewood Cliffs
    NJ: Prentice-Hall 1959.
MORGAN, L.V.: Origin of attached X-chromosomes in Drosophila
    melanogaster and the occurrence of non-disjunction of X's
    in the male. American Naturalist 72, 434-446 (1938).

3.5 Genetic Localization of Mutations Within the Genome

A central problem in genetic investigations is the analysis
of new, unknown mutations. Initially one assumes a new mutation when a mutant phenotype appears. In crosses we determine first whether the mutant phenotype is heritable or whether it is a phenocopy or developmental defect. If the mutant character proves to be a simple Mendelian factor, one
tries to characterize it with respect to dominance, viability, fertility, etc. The next step of the analysis is to attempt to localize the gene within the genome: One determines
the linkage group (i.e. the chromosome) to which it belongs
and then where it lies on that chromosome. The localization
(map position) is established with the help of recombination
experiments. In Drosophila these problems can be solved with
comparatively simple crosses using suitable genetic markers.

Since genes are located on chromosomes they follow the behavior of chromosomes in cell division. Homologous chromosomes
segregate from each other during meiosis. Different genes
located on the same chromosome tend to be inherited together,
i.e. they form a linkage group. The segregation of one pair
of homologs does not influence the segregation of a different
pair, therefore a pair of alleles located on one particular
chromosome pair will assort independently of a pair belonging
to a different linkage group.

The assignment of a particular linkage group to a particular
chromosome recognizable cytologically involves the study of
the phenotypic effects of genes together with observable
structural differences in chromosomes (cytological studies).
Once some genes have been assigned to a particular chromosome, others can be assigned to the same chromosome by demon-
strating linkage to a gene of a known linkage group. For
example, if we know that gene a is on chromosome 2 and can
show that gene b is linked to gene a, then we can deduce that
gene b is also on chromosome 2. Since crossing over separates linked genes, such determinations depend on the genes
being close enough together to demonstrate linkage statistically unless crossing over can be suppressed.

In Drosophila there are special marker lines available which
allow the determination of the linkage group of any gene.
One such line is the Cy/Pm; H/Sb line. Cy (Curly wings,
2-6.1) and Pm (Plum eye color, 2-104.5) are both dominant
markers located on the 2nd chromosome. Both are lethal in
homozygotes and in this strain are found on chromosomes carrying large complex inversions which effectively eliminate
the recovery of chromatids which have resulted from crossing
over in heterozygotes (balancer chromosomes, see experiment
7.2). H (Hairless, 3-69.5) and Sb (Stubble, 3-58.2) are also
homozygous lethals and associated with inversions. The complete genotype of the Cy/Pm; H/Sb line for the 2nd and 3rd
chromosomes is:

$$\frac{\text{Cy} \qquad +}{+ \qquad \text{Pm}} \qquad \frac{\text{H} \qquad +}{+ \qquad \text{Sb}}$$

Since the effects of crossing over between these chromosomes
are eliminated, the Cy chromosome always segregates from the
Pm, and H segregates from Sb. Independent assortment is normal so that the gametes formed have all combinations of 2nd
and 3rd chromosomes (Cy with H, Cy with Sb, Pm with H, and Pm
with Sb).

This line is used to cross with a homozygous line of a gene
of unknown linkage group, and F_1 individuals are test
crossed. The dominant markers show the presence of particular 2nd and 3rd chromosomes, and the test cross progeny obtained will demonstrate segregation of the unknown gene from
a marker gene belonging to the same linkage group and independent assortment with a marker belonging to a different
linkage group. In the test cross results, if two markers are
segregating from each other, every phenotypic class will have
either the one or the other. No class will have both, and no
class will have neither. If two marker genes are assorting
independently, all combinations are possible (both mutants,
one mutant only, the other mutant only, neither mutant).
In practice it is easy to assign genes located on the X chromosome to the correct linkage group because of the unique
pattern of sex-linked inheritance (see experiment 3.2); so
the only problem arises with autosomal genes. By using mar-

ker genes on chromosomes 2 and 3, the position of any gene
may be determined: A gene on 2 will segregate from the 2nd
chromosome marker and assort independently with the 3rd chromosome marker; a gene on 3 will segregate from the marker on
3 and assort independently with the marker on 2; and a gene
on 4 will assort independently with both markers.

Problem

Localize an unknown recessive mutation on one of the three
autosomes.

Material

Virgin females of an unknown mutant and males of the Cy/Pm;
H/Sb strain are needed.
Note to instructors: Each student may be given a different
autosomal recessive mutant with an easily identifiable phenotype (e.g. b, e, se, spa^{pol}, vg, etc.).

Experimental work and interpretation

Day 1 (week 1): Examine and record the phenotype of your
unknown mutant and the phenotypes of the four dominant markers.
Cross the females of your unknown line with the Cy/Pm; H/Sb
males. Prepare two vials with several pairs of flies in each.
Day 4 (week 2): Remove the parents from your cross vials and
examine them to refresh your memory as to the phenotype of
your mutant.
Day 12 (week 3): You are going to use one of the four different phenotypic classes expected in the F_1 as the male
parent for the test cross. Your choice of class should be
such that the phenotype of the marker genes chosen will not
interfere with the classification of your unknown in the next
generation. For example, if your unknown was an eye color,
do not choose a class carrying Pm, if it was a wing characteristic, do not choose a class carrying Cy.

Anesthetize the F_1 being careful not to kill the flies.
Choose 4-6 males of the class you wish to use as parents for
the test cross and transfer them to an empty vial. Record
the phenotype used.

Phenotype of males for test cross: ....

Examine the F_1 progeny and record the four classes present:

Class	2nd chromosome marker	3rd chromosome marker
1		
2		
3		
4		

Make the test cross by crossing virgin females of your unknown mutant line with the F_1 males of the class chosen.
Set up two vials with several pairs of flies in each. Using
u as a symbol for your unknown mutant, write the genotype of
the cross made for all three genes involved (your unknown,
the chosen 2nd chromosome marker and the chosen 3rd chromosome marker).

.... x ....

Day 15 (week 4): Discard the parents from your cross vials.
Day 23 (week 5): Classify the test cross progeny and construct a table showing the different phenotypic classes obtained.

<u>Questions</u>:

(a) Did your unknown mutant segregate from the 2nd chromosome marker?

(b) Did your unknown mutant segregate from the 3rd chromosome marker?

(c) Did your unknown mutant assort independently of the marker on chromosome 2?

(d) Did your unknown mutant assort independently of the marker on chromosome 3?

(e) To which linkage group does your unknown gene belong?

<u>Literature</u>

MORGAN, T.H.: Random segregation versus coupling in Mendelian inheritance. Science $\underline{34}$, 384 (1911). In: Papers on Genetics. A Book of Readings (ed. L. Levine). St. Louis MO: Mosby Comp. 1971.

3.6 Mapping on a Chromosome

Gene mapping is based on the principle that genes located on the same chromosome are separated from each other by crossing over during meiosis, and furthermore that the frequency with which crossing over occurs in a particular region of the chromosome is a function of the length of that region. The further apart two genes are, the higher the probability that they will be separated by crossing over. The unit of distance on genetic maps is one per cent recombination (one centimorgan). If two genes are 15 map units apart, this means that a female fly heterozygous for the two genes on average will produce eggs of which 85% will have the two genes in the same arrangement on the two chromosomes as they were in the cells of the female (the parental types) and 15% will have a new arrangement (the recombinant types).

For example, if a female heterozygous for two hypothetical linked genes a and b had one chromosome carrying the two mutant alleles and the other homolog carrying the two wild type alleles, we could represent the chromosomes as follows:

$$\frac{\underline{\quad a \quad\quad b \quad}}{+ \quad\quad +}$$

After meiosis, the chromosomes passing into gametes would be of two types and each type contains two complementary classes:

85% parental types (no recombination between a and b)	42.5%	a b
	42.5%	+ +
15% crossover types (recombination between a and b)	7.5%	a +
	7.5%	+ b

The arrangement of the genes shown above, where the two mutants are on one chromosome and the two wild type alleles on the other, is known as the coupling or cis configuration. A heterozygote may also have the genes arranged with one mutant on the same chromosome as the wild type allele of the

other gene. This is known as the repulsion or trans configuration. With the genes arranged in that way, the classes a +
and + b would be the parental types and a b and + + the recombination classes.

The three point test

When a test cross involving three or more linked genes can be
made, then information about the order of the genes as well
as their distances from each other can be obtained.
Suppose that we had information from a test cross of a triple
heterozygote but did not know the order of the three genes or
how they were arranged on the chromosomes. With three genes,
we will have four different kinds of classes with respect to
recombination: Two classes will be the parental types.
These will always be the largest classes. Another pair of
complementary classes will be formed by gametes from the heterozygote that have undergone a single crossing over between
the left gene and the middle one, and the third type will be
the two classes resulting from single crossing over between
the middle gene and the right one. The fourth type expected
would result from chromosomes in which there had been crossing over in both regions under study, the so-called double
crossover classes. These will normally be the smallest classes. The single crossover classes will be intermediate in
size between the parental classes and the double crossover
classes. This information allows us to determine the arrangement of the three genes and their order on the chromosome.

An example will show how this is done: Since the phenotype
of each class in the test cross progeny is the same as the
genotype of one product of meiosis from the heterozygous parent and two homologous chromosomes participate in crossing
over at meiosis, the first step is to group the classes into
complementary pairs and sum their totals.

<u>Test cross data</u>

<u>Original data</u>		<u>Grouped data</u>		
Phenotypic class	No. of indiv.	Phenotypic class	No. of indiv.	Pairs of classes
$a^+b^+c^+$	5	$a^+b^+c^+$	5	
a^+b^+c	183	$a\ b\ c$	3	8 smallest
$a^+b\ c^+$	48	a^+b^+c	183	
$a^+b\ c$	19	$a\ b\ c^+$	175	358 largest
$a\ b^+c^+$	23	$a^+b\ c^+$	48	
$a\ b^+c$	44	$a\ b^+c$	44	92 intermed.
$a\ b\ c^+$	175	$a\ b^+c^+$	23	
$a\ b\ c$	3	$a^+b\ c$	19	42 intermed.
Total	500			

Now, knowing that the largest pair must represent the parental types, we can see that a^+, b^+ and c must have been present on one homolog and a, b and c^+ on the other, but the order is still uncertain. There are three possibilities:

$$\frac{a^+b^+c}{a\ b\ c^+} \qquad or \qquad \frac{a^+c\ b^+}{a\ c^+b} \qquad or \qquad \frac{b^+a^+c}{b\ a\ c^+}$$

Knowing that the two smallest classes are $a^+b^+c^+$ and a b c and that these must have undergone double crossing over, we can work out that there is only one of the three possible arrangements that will produce the two smallest classes by double crossing over. Since exchanges both to the left and to the right of the middle gene result in the movement of the middle gene from the one chromosome to the other, we can deduce the correct order by comparing the parental classes with one of the double crossover classes. One of the parental classes will always be identical to one double crossover class so far as two of its genes are concerned. These must be the end genes, and the third gene which has a different allele must be the middle locus.

From our example, the parental classes had a^+, b^+ and c on one homolog and a, b and c^+ on the other. Comparing one double crossover class a b c to these, we can see that it is identical to a b c^+ for the alleles of a and b but carries a different allele of c. From this we deduce that the c locus must be in the middle between a and b. This is the only arrangement of the three shown above that will produce the double crossover classes by exchanges in the two regions.

Now we rewrite the genes in the correct order and identify the type of events that gave rise to each pair of complementary classes.

The parental chromosomes were:

$$\text{a } c^+\text{b}$$
$$\text{===============}$$
$$\text{a}^+\text{c b}^+$$

Single crossing over between the a and c loci will give:

$$\text{a } c^+\text{b} \quad\searrow\quad \text{a c b}^+$$
$$\text{a}^+\text{c b}^+ \quad\nearrow\quad \text{a}^+\text{c}^+\text{b}$$

Single crossing over between the c and b loci will give:

$$\text{a } c^+\text{b} \quad\nearrow\quad \text{a } c^+\text{b}^+$$
$$\text{a}^+\text{c b}^+ \quad\searrow\quad \text{a}^+\text{c b}$$

Double crossing over will give:

$$\text{a } c^+\text{b} \quad\nearrow\quad \text{a c b}$$
$$\text{a}^+\text{c b}^+ \quad\searrow\quad \text{a}^+\text{c}^+\text{b}^+$$

Now the map distances can be worked out. Since one map unit equals one % recombination, the distance from a to c is the sum of all the classes which have had recombination between a and c divided by the total and expressed as a percentage.

$$\text{Recombination a to c} = \frac{\text{single c.o. (a-c) + double c.o.}}{\text{total}}(100)$$

$$= \frac{92 + 8}{500}(100) = 20$$

Recombination c to b = $\dfrac{\text{single c.o. (c-b) + double c.o.}}{\text{total}}$ (100)

$$= \frac{42 + 8}{500}(100) \quad = \quad 10$$

We can now construct the chromosome map for this region of the chromosome. Note that only the locus designations and not particular alleles are shown on the map.

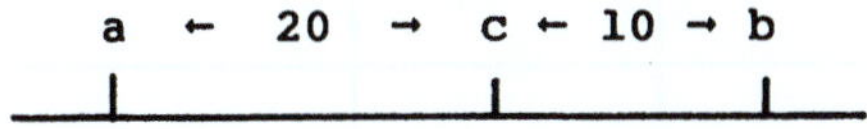

Problem

Construct a gene map for three sex-linked genes.

Material

Virgin females carrying the three markers w (white eyes), m (miniature wings) and f (forked bristles). Males of a wild type strain.

Experimental work and interpretation

<u>Day 1 (week 1)</u>: Cross virgin females of the strain w m f with w^+ m^+ f^+ males in a bottle.
<u>Day 4 (week 2)</u>: Discard the parents from the bottle.
<u>Day 12 (week 3)</u>: Examine the phenotypes of the F_1 generation. The females are all heterozygous for the three genes, and the males are hemizygous for the three recessive genes. For this particular type of cross with sex-linked genes, the F_1 intercrossed give a test cross. Transfer about 10 females and 10 males from the F_1 to a fresh bottle.
<u>Day 15 (week 4)</u>: Discard the flies from the bottle.
<u>Day 23 (week 5)</u>: Classify the test cross progeny into the 8 expected phenotypic classes. Fill in your results in the table on the next page.

<u>Results</u>

Class	Phenotype Eye Wing Bristle	Number of flies	Type of event
1	w m f		
2	+ + +		
3	w + +		
4	+ m f		
5	w m +		
6	+ + f		
7	w + f		
8	+ m +		
	Total:		

In this experiment the parental classes are known because of
the way in which the cross was made.
Were the parental classes the largest? ....
Can you confirm that the gene order is correct as given? ....
In the table above, fill in the type of event that occurred
to produce the maternal X chromosomes of the individuals in
that class (no c.o., single c.o. w-m, single c.o. m-f, double
c.o.). Use the data from the table to construct a chromosome
map for the three genes.

<u>Literature</u>

STURTEVANT, A.H.: The linear arrangement of six sex-linked
    factors in Drosophila, as shown by their mode of associa-
    tion. J. Exp. Zool. <u>14</u>, 43-59 (1913). In: Classic
    Papers in Genetics (ed. J.A. Peters). Englewood Cliffs
    NJ: Prentice-Hall 1959.

3.7 Segregation of Compound Chromosomes

Translocations involving whole chromosome arms are extreme
cases of chromosome aberrations. This experiment will demonstrate the specific consequences of such aberrations when
whole arms of homologous chromosomes have been exchanged.
The diagram below shows the generation of compound chromosomes of the second chromosome in Drosophila. In the diagram
L stands for the left or short arm and R for the right arm.
The circle represents the centromere and the thick regions
centromeric heterochromatin where the breaks occurred.

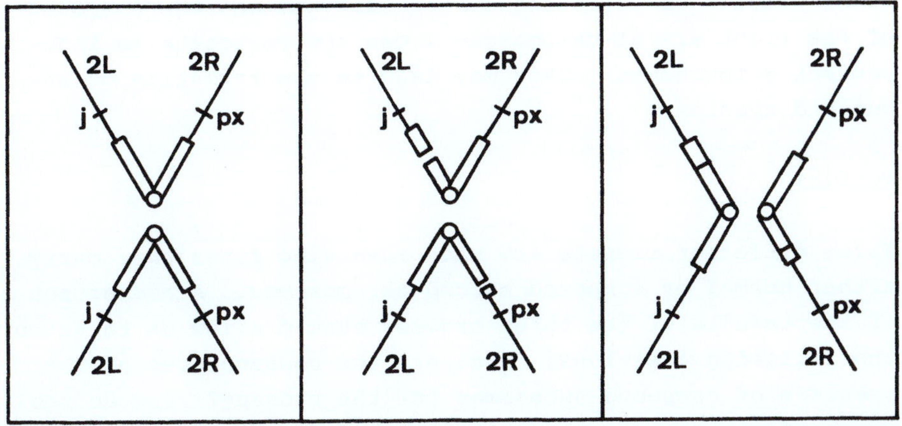

original homologous   chromosomes break   broken ends rejoin
chromosomes   near the centromeres   producing translo-
                                                   cation

In a diploid cell the two second chromosomes carry the marker
j (=jaunty) on the left arm and the marker px (=plexus) on
the right arm. A chromosome break is induced in the heterochromatin to the left of the centromere of one chromosome.
In the homologous chromosome a corresponding break is induced
in the heterochromatin to the right of the centromere. The
broken ends now rejoin exchanging. In this way two new chromosomes are created, so-called compound chromosomes. They
carry two homologous chromosome arms attached to one centromere. The two arms are attached in a mirror image way;
therefore such a compound chromosome is called a "reversed
metacentric" (RM). The genetic description of the one com-

92

pound chromosome shown above is: C(2L)RM, j. The symbols
mean: C = compound, 2L = left arm of chromosome 2, RM = reversed metacentric. After the comma the genetic markers are
given (here j = jaunty). In an analogous way the compound
chromosome with the marker plexus is designated as C(2R)RM,
px. It is important to realize that the diploid chromosomal
information content of the cell has not been altered by this
specific type of chromosome rearrangement. No chromosomal
material has been lost; therefore the cell remains viable.
Only the linkage conditions have been changed. Serious consequences of such chromosome aberrations will only appear
when the chromosomes segregate in meiosis and fertilization
takes place. The compound of the left arm and the compound
of the right arm of chromosome 2 can now segregate as independent chromosomes. This may lead to the formation of aneuploid zygotes.

Problem

Three different crosses are made involving flies that carry
either normal or compound second chromosomes. A comparison
of the results of the three crosses should allow us to answer
the following questions: What are the consequences of the
presence of compound autosomes for the progeny? How do the
compound chromosomes segregate in the two sexes?

Material

For the three crosses virgin females and males are collected
from the following strains:
    1)   wild type
    2)   C(2L)RM, j ; C(2R)RM, px
    3)   C(2L)RM, b ; C(2R)RM, cn.
The markers used are:
    jaunty (j, 2-48.7)   curved wings
    plexus (p, 2-100.5)   additional veins on wing
    black (b, 2-48.5)   black body color
    cinnabar (cn, 2-57.5)   light red eye color

Experimental work and analysis

As a preparation for the experiment we first have to work out which types of gametes are produced when a compound autosome is present. In the following diagram, enter the types of oocytes a female with the genotype shown will produce:

Genotype:

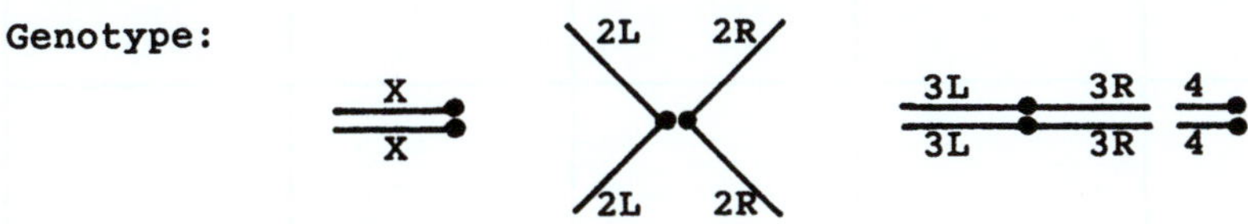

Expected oocytes:

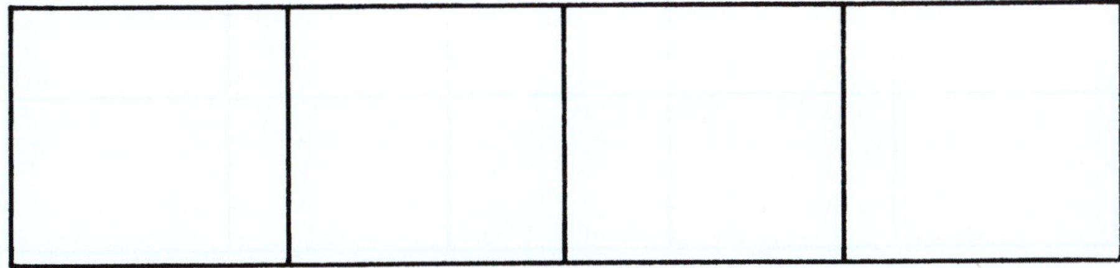

Give reasons for this expectation.

The three different crosses to be made together with the checkerboards are given below. Since all other chromosomes are normal, we need consider only the second chromosomes.

Cross 1

♀ C(2L)RM, j; C(2R)RM, px   x   ♂ wild type

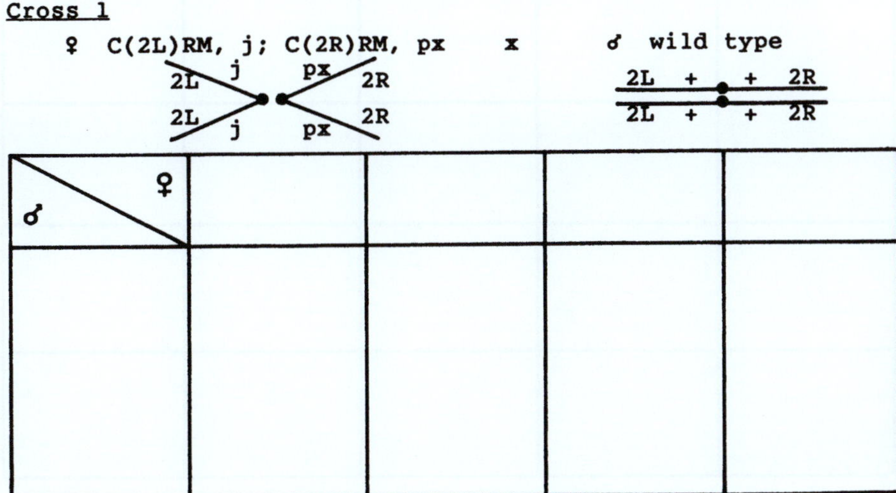

Cross 2

♀ C(2L)RM, j; C(2R)RM, px   x   ♂ C(2L)RM, b; C(2R)RM, cn

♂ \ ♀				

Cross 3

For this cross we collect progeny from cross 2.

♀ C(2L)RM, b; C(2R)RM, px   x   ♂ C(2L)RM, j; C(2R)RM, cn

♂ \ ♀				

In the following table, list all the phenotypes of the progeny which are expected in each cross together with their expected relative frequencies.

To make the crosses, establish the cultures as follows:

<u>Day 1 (week 1)</u>: Two virgin females and two males are placed in each vial.

<u>Day 4 (week 2)</u>: Discard parents.

<u>Day 12 (week 3)</u>: Classify and count the progeny. Collect the appropriate virgin females and the males for cross 3 from the progeny and transfer to fresh vials. Enter the phenotypes found and their frequencies in the table below. Do all the expected phenotypes occur? What are possible reasons for discrepancies between expectation and observed frequencies? How do the compound chromosomes segregate in the two sexes?

Cross	Phenot.:	+ +	j px	j cn	b px	b cn
1	expected					
1	observed					
2	expected					
2	observed					
3	expected					
3	observed					

<u>Literature</u>

HOLM, D.G.: Compound Autosomes. In: The Genetics and
    Biology of Drosophila, Vol. 1b (eds. M. Ashburner and E.
    Novitski). London: Academic Press 1976.
HILLIKER, A.J., APPELS, R., SCHALET, A.: The genetic ana-
    lysis of <u>D</u>. <u>melanogaster</u> heterochromatin. Cell <u>21</u>,
    607-619 (1980).

3.8 Meiotic Mutants

Initially the sequence of events during the meiotic divisions
was interpreted based on observations with the light microscope. Today it is possible to investigate the meiotic processes with electron microscopy and with biochemical methods. Due to the fact that the essential processes of meiosis
are also under genetic control, the so-called meiotic mutants
help to analyze and understand various aspects of meiosis.
After the discovery of the first meiotic mutant in Drosophila
in 1933 such mutants were also discovered in various other
organisms: yeast, Neurospora and other lower eukaryotes,
higher plants (e.g. maize), as well as a large number of new
mutants in Drosophila (Baker et al. 1976). Meiotic mutants
of <u>Drosophila</u> <u>melanogaster</u> exhibit characteristic disturbances in specific steps of meiosis. These disturbances lead
in most cases to reduced fertility and/or viability of the
affected flies. Using a mutant in which female meiosis is
disturbed we are able to analyze two aspects: 1) The influence on meiotic chromosome segregation, which is expressed in
altered disjunction conditions, and 2) the influence on meiotic crossing over. For this purpose, in specific crosses we
can observe the frequencies of nondisjunction of the X chromosomes as well as the rates of crossing over between two
distant X-chromosomal markers. In parallel crosses we compare the same phenomena in wild type females and in females
with the mutation mei-9. The mutation mei-9 is recessive and
located on the X-chromosome at locus 6.0. Of the several
alleles known, we will use the two alleles mei-9^a and
mei-9^{L1}. Biochemical investigations have shown that the
mei-9 mutants not only exhibit disturbances in meiosis but
also possess an increased sensitivity to the toxic and mutagenic effects of several chemical mutagens. This increased
sensitivity is caused by a defect in cellular DNA excision
repair.

<u>Material</u>

The following five strains are needed:
1) wild type   4) $y\ w^a\ mei\text{-}9^a$
2) B   5) $y\ mei\text{-}9^{L1}\ cv\ /\ y^+\ Y\ B^S$
3) $mei\text{-}9^{L1}$

<u>Problem 1: Influence of the mei-9 mutation on nondisjunction frequencies</u>

We wish to determine the frequencies of the exceptional phenotypes caused by nondisjunction of the X chromosomes among
the progeny of normal wild type females and among the progeny
of mei-9 females. In order to have comparable situations,
the two types of females are crossed with the same males.
The X chromosomes of the mei-9 females carry the genetic
markers y (yellow, 1-0.0; yellow body color), w^a (white-
apricot, 1-1.5; apricot eye color) and the mutation $mei\text{-}9^a$
(meiotic-9, 1-6.0; disturbances in meiosis and defects in DNA
excision repair). The tester males carry an X chromosome
with the dominant marker B (Bar, 1-57.0; Bar eyes). The different Bar phenotypes have been explained in experiment 3.4.
Fill in the gametes and the expected progeny genotypes in the
checkerboards below.

<u>Cross A:</u>

       ♀ wild type   x   ♂ X, B / Y

		nondisjunction (ND)	
♂ \ ♀			

Cross B:

$♀\ y\ w^a\ mei\text{-}9^a\ /\ y\ w^a\ mei\text{-}9^a\qquad x\qquad ♂\ X,\ B\ /\ Y$

		nondisjunction (ND)	
♂ \ ♀			

Record the results of your experiment below.

Results

Series	XX ♀	XY ♂	XXY ♀	X0 ♂	Total	ND (%)
wild type						
mei-9						

The frequency of nondisjunction is calculated as:

ND (%) = (XXY + X0) . 100 / (XX + XY + XXY + X0)

Conclusions

What effect did the mei-9 mutant have on nondisjunction?

Problem 2: Influence of the mei-9 mutation on crossing over frequencies

In this experiment we determine the frequencies of crossing over between the marker genes y (yellow, 1-0.0; yellow body color) and cv (crossveinless, 1-13.7; wings without cross-veins). In order to establish the influence of the mei-9 mutation, we perform the experiment with wild type and mei-9 females in parallel. The females in which we want to record the frequencies of crossing over have to be heterozygous for the markers y and cv on the one hand and heterozygous or ho-mozygous, respectively, for mei-9 on the other hand. To achieve this we use the following males for both crosses: y mei-9^{L1} cv / y$^+$ Y B^S. The marked Y chromosome (with the markers y$^+$ = wild type allele of yellow and B^S = an extreme Bar allele) allows the recognition of XXY females that might appear in the F$_1$ by their reduced eye size. These females have to be excluded from further use because they produce abnormal progeny.

First examine the schematic cross diagramed below. The sym-bol 9 is used as an abbreviation for mei-9^{L1}. Now work out the phenotypes of the four male classes in the F$_2$ (indepen-dent of the mutation mei-9). The recombination frequencies can be determined directly with the F$_2$ males because they are hemizygous for the X chromosomes derived from female meiosis (with crossing over). The F$_2$ females are all wild type because in the cross with the F$_1$ females normal wild type males are used.

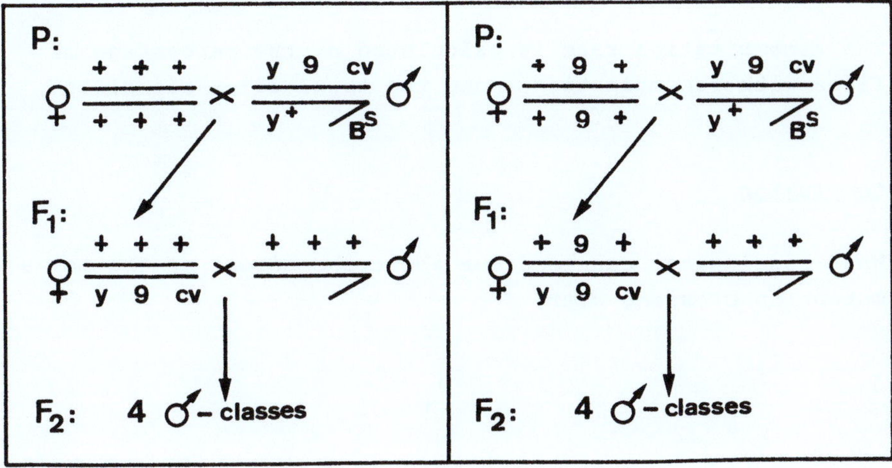

<u>F$_1$ cross</u>:

	♀	without recombination		with recombination	
♂					
⟍					
Phenotype: body color crossveins					

Record the results obtained for the F$_2$ in the table below.

<u>Results</u>:

Series	y$^+$ cv$^+$	y cv	y$^+$ cv	y cv$^+$	Total	Rec. (%)
wild type						
mei-9						

The recombination rate is calculated as the percentage of flies with recombination among the total flies classified.

<u>Conclusion</u>

What conclusions can be drawn about the effect of the mei-9 mutant on crossing over?

<u>Literature</u>

BAKER, B.S., CARPENTER, A.T.C., ESPOSITO, M.S., ESPOSITO, R.E., SANDLER, L.: The genetic control of meiosis. Ann. Rev. Genetics <u>10</u>, 53-134 (1976).
BOYD, J.B., HARRIS, P.V.: Mutants partially defective in excision repair at five autosomal loci in <u>Drosophila</u> <u>melanogaster</u>. Chromosoma (Berl.) <u>82</u>, 249-257 (1981).
BOYD, J.B., MASON, J.M., YAMAMOTO, A.H., BRODBERG, R.K., BANGA, S.S., SAKAGUCHI, K.: A genetic and molecular analysis of DNA repair in Drosophila. J. Cell Sci. Suppl. <u>6</u>, 39-60 (1987).
GENEROSO, W.M., SHELBY, M.D., DE SERRES, F.J. (eds.): DNA Repair and Mutagenesis in Eukaryotes, Vol. 15 of Basic Life Sciences. New York, London: Plenum Press 1980.
HARRIS, P.V., BOYD, J.B.: Excision repair in Drosophila. Analysis of strand breaks appearing in DNA of mei-9 mutants following mutagen treatment. Biochim. Biophys. Acta <u>610</u>, 116-129 (1980).

4. Phenogenetics

4.1 Temperature Effect on Expression of Phenotype

In formal genetics alternative phenotypic differences between different genotypes are analyzed, and it is assumed that there is a direct relationship between a particular genotype and the corresponding phenotypic trait. In many cases no such simple direct correlation is found. In the branch of genetics known as phenogenetics, the expression of genes is analyzed in order to gain more information about how genes produce their phenotypic effects. Ultimately gene-effect relationships should be described in terms of chemical reactions, but before this can be done, the structural or functional traits to be studied must be investigated and characterized by classical genetic methods. When experiments are designed to investigate the effects of genotype and environment on the expression of particular genes, the results show that genes do not usually function independently of each other and of the environment but represent a complex gene-effect network.

Certain terms are used to describe particular aspects of gene expression:

<u>Penetrance</u>: In some cases not all of the individuals having a genotype that would be expected to give a mutant phenotype show the expected trait. Penetrance is a measure of the proportion showing the expected phenotype.

$$\text{Penetrance} = \frac{\text{number of individuals with mutant phenotype}}{\text{number of individuals with specified genotype}}$$

<u>Expressivity</u>: Among individuals expressing a particular mutant phenotype, there may be variation in the severity of the deviation from the normal. Note: Normal individuals are excluded from the calculation of the expressivity.

Expressivity = class mean of mutant individuals

<u>Phenocopy</u>: Certain environmental effects may result in a phenotype that is abnormal in an individual with normal genotype. This is known as a phenocopy.

Differences in penetrance and expressivity can have various causes:

<u>Influence of the genetic background</u>: Since genes work bio-chemically it is not surprising that the effect of one gene locus may be affected by the action of many other genes. For convenience, genes having a major identifiable effect are called <u>oligogenes</u>. In cases where so many genes affect a particular characteristic such as quantitative traits (size, etc.) that the action of any single gene has less effect than the environment, we speak of <u>polygenes</u>. Genes having a major effect on the expression of other nonallelic genes are known as <u>modifier genes</u>. A particular genotype at one locus may also produce differing phenotypes depending on differences in the genetic milieu in genes influencing dominance, position effects, presence of mutable genes, aneuploidy, polyploidy.

<u>Influence of the environment</u>: Environmental factors such as nutrition, temperature, humidity, light, etc. can have an effect on the phenotypic expression of a genotype. Some traits are stable in any environment while others are environmentally labile. This is studied by analyzing different genotypes in a stable environment and one genotype in different environments.

<u>Problem</u>

What is the effect of an increase in culture temperature from the standard 25°C to 29.5°C on the expression of vestigial wings in vg/vg flies?

<u>Material</u>

A well fed culture of vestigial flies grown at 25°C. An incubator adjusted to 29.5°C.

<u>Procedure</u>

(1) Large numbers of well fed vg/vg flies are put into standard culture bottles for 3 hours of egg laying and then removed.

vestigial wings

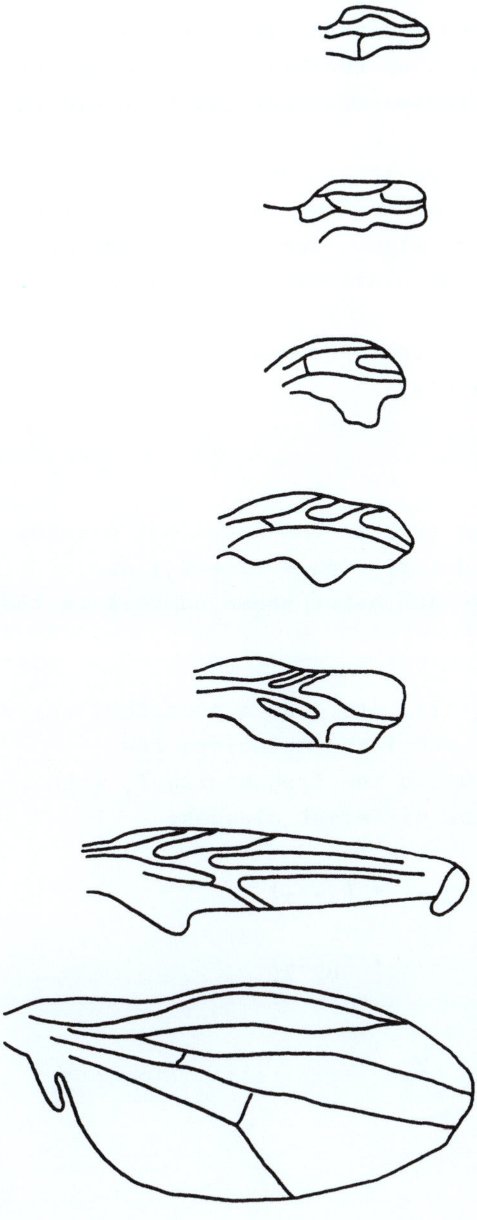

	25°C	29.5°C
VII		
VI		
V		
IV		
III		
II		
I		
N		

(2) At the end of the egg laying period some of the culture
bottles are placed in a 29.5°C incubator; the others are kept
at 25°C.

(3) The bottles at the higher temperature are kept in the
29.5°C incubator for <u>100 hours</u> and then placed at 25°C.
Note: After 100 hours there is no further influence on the
phenotype but the mortality increases drastically at higher
culture temperature.

(4) Approximately 10 to 12 days after the start of the experiment, the surviving flies of the two series are sorted
into classes of different wing size. Use the drawings of
wings shown in the table on the previous page to define the
means of the classes.

(5) Using the t test check whether the mean wing size in the
two series differs significantly.

<u>Analysis</u>

Starting with the frequencies (f_i) in the different classes
determine the total number ($N = f_1 + f_2 + .. + f_7$) of
wings analyzed in each series and enter these numbers in the
table.

Use the class number (I, II, III, etc.) as a numerical value
x_i ($I = x_1 = 1$; $II = x_2 = 2$; etc.) and calculate the
mean ($\bar{x}$) for the two series using the frequencies f_i with
which wings are present in the different classes:

$$\bar{x}_A = (1/N_A) \cdot (f_{A1}x_1 + f_{A2}x_2 + \ldots + f_{A7}x_7)$$

$$\bar{x}_B = (1/N_B) \cdot (f_{B1}x_1 + f_{B2}x_2 + \ldots + f_{B7}x_7)$$

$$\bar{x}_A = \ldots\ldots\ldots\ldots\ldots \qquad \bar{x}_B = \ldots\ldots\ldots\ldots\ldots$$

To test the difference between the means of the two series
$\bar{x}_A - \bar{x}_B$ calculate the deviation according to the following formula (Z = sum over all classes 1 to 7):

$$S^2 = [1/(N_A+N_B-2)] \cdot [Z(x_i-\bar{x}_A)^2 + Z(x_i-\bar{x}_B)^2]$$

$$= \ldots\ldots\ldots\ldots$$

From S^2 we can calculate the standard deviation:

$$S = \sqrt{S^2} = \ldots\ldots\ldots\ldots$$

Now determine the t value as follows:

$$t = (\bar{x}_A-\bar{x}_B)/S \cdot \sqrt{(N_A \cdot N_B)/(N_A+N_B)} = \ldots\ldots\ldots$$

The number of degrees of freedom is:

$$d.f. = N_A + N_B - 2.$$

In the table of t values below, read the t value corresponding to d.f. at P = 0.05 (i.e. error probability 5%) and
compare it with the t value calculated as above. If the calculated value is larger than the value in the table, then the
two series A and B are statistically significantly different
from each other.

t values at P = 0.05

d.f.	t		d.f.	t		d.f.	t
20	2.086		90	1.987		200	1.972
30	2.042		100	1.984		300	1.968
40	2.021		120	1.980		400	1.966
50	2.009		140	1.977		500	1.965
60	2.000		160	1.975		1000	1.962
70	1.994		180	1.973		∞	1.960
80	1.990						

Conclusions

What conclusion can you draw about the effect of temperature
on the expression of the vg gene? Was the penetrance affected? Was the expressivity affected?

Literature

HARNLY, M.H.: The temperature effective periods and the
     growth curves for length and area of the vestigial wings
     of _Drosophila melanogaster_. Genetics _21_, 84-103 (1936).
LI, J., TSUI, Y.: The development of vestigial wings
     under high temperature in _Drosophila melanogaster_. Ge-
     netics _21_, 248-263 (1936).
STANLEY, W.F.: The effect of temperature on vestigial wing
     in _Drosophila melanogaster_, with temperature effective
     periods. Physiol. Zoology _4_, 394-408 (1931).
STANLEY, W.F.: The effect of temperature upon wing size in
     Drosophila. J. Exp. Zoology _69_, 459-495 (1934).
WARDLAW, A.C.: Practical Statistics for Experimental Biolo-
     gists. Chichester UK: John Wiley and Sons 1985.

4.2 Temperature Effect on a Homeotic Mutant

In the development of any complex organism from a single
cell, there is a specific sequence of events occurring that
controls the expression of a group of genes each of which
must act at the correct time and in the correct place to give
normal development.

Groups of genes must be switched on and off. How this regulation occurs is a field of intense research activity at present. Embryology and developmental biology are going through
a phase of stimulation by applying the techniques of molecular biology which have allowed the study of genes at the level of their DNA sequences, the identification of primary gene
products and their localization in developing embryos and
immature stages.

A number of Drosophila homeotic mutants have been known for
some time which upset normal morphological development by
transforming a structure typical of one body part into one
typical of another part (e.g. turning an antenna into a
leg). Another class of mutants upset the normal segmentation
pattern of the animal. When several of these genes were
cloned and their nucleotide sequences determined, one region
was found which was very similar in the different genes.
This so-called "homeobox" is a highly conserved sequence of
nucleotides which code for about 60 amino acids. A homeobox
has since been identified as a part of many genes that show
temporal or spatial expression in organisms as diverse as the
frog, the mouse and man.

The precise function of the homeobox region is not known at
present, but one possibility is that products of genes with
homeoboxes may act as DNA-binding transactive regulators of
the activity of other genes.

The major homeotic genes of Drosophila fall into two main
gene complexes, the antennapedia complex and the bithorax
complex, both of which are located on the third chromosome.
One of the most striking of these mutants is ss[a] (spineless

aristapedia, 3-58.3) which causes the transformation of a head structure, the arista of the antenna, into a thoracic structure, the tarsus of a leg. One allele of this gene, ss^{a40a}, shows a temperature effect.

Problem

What effect does a high developmental temperature have on the expression of ss^{a40a}?

Material

A well-fed culture of ss^{a40a} kept at 25°C. An incubator at 29°C.

Experimental procedure

Well fed flies homozygous for ss^{a40a} are put into standard culture bottles for 3 hours of egg laying and then removed. At the end of the egg laying period, some of the bottles are placed in a 29°C incubator and the rest kept at 25°C. Approximately 9 to 12 days after the start of the experiment, the surviving flies of the two series are classified for antenna form. The results are recorded in the table below.

Results

Temp.	Both antennae wild type	Intermediate	Both antennae strongly transformed
25°C			
29°C			

Conclusions

Does temperature affect the penetrance and/or expressivity of
this gene?
At which temperature does the product of this gene behave
most like the wild type product?
What experiment could be done to find out if the effect of
temperature is continuous, i.e. becoming more abnormal with
increasing temperature change?

Literature

GRIGLIATTI, T., SUZUKI, D.: Temperature-sensitive mutations
     in Drosophila, VIII. The homeotic mutant ss^{a40a}.
     Proc. Nat. Acad. Sci. (USA) <u>68</u>, 1307-1311 (1971).

4.3 Mutants with Abnormal Eye Color

Mutants of <u>Drosophila melanogaster</u> that show an abnormal eye
color can be quite conspicuous. Due to the fact that defects
in the biosynthesis of eye pigments have very little effect
on the viability of the carriers, such mutants have been discovered, isolated and analyzed in large numbers. An overview
of the various types of eye color mutants shows that they can
be grouped phenomenologically as follows:

(1) brick red: Wild type flies and heterozygous carriers
                  of recessive eye color alleles. This eye
                  color is designated "normal".

(2) brown: A number of mutants show more or less
                  intensive brown eyes.

(3) bright red: A number of mutants exhibit more or less
                  bright red eyes.

(4) white: In some strains the eye color is yellowish
                  white to pure white.

<u>Problem</u>

How many different biosynthetic pathways lead to the eye pigments which produce the brick red eye color in the wild type?

The simplest hypothesis is that in the wild type there are
basically two different types of pigments present: bright
red and brown. The following three questions will lead to an
experimental approach by which the problem can be tackled:

(a) Flies which are homozygous bw/bw show a brown eye col-
or. Which type of pigment is most probably not produced in
these animals?

(b) Which type of pigment is not produced in flies with
bright red scarlet eyes (st/st)?

(c) The genes bw and st are in two different linkage
groups. What types of progeny (eye color) are expected in
the F_2 starting with an initial cross of brown-eyed with
bright red-eyed flies?

<u>Material</u>

Virgin bw/bw females and st/st males. The gene brown (bw)
maps at chromosomal position 2-104.5 and the gene scarlet
(st) at 3-44.0.

<u>Experimental work and evaluation</u>

Fill in the two checkerboards on the next page and for each
genotype, predict the expected phenotype (eye color) based on
the hypotheses given above. Enter the eye color in the space
at the bottom of each square in the checkerboards. Remember
always to write the genetic formulas completely, i.e. always
give both alleles at the brown as well as at the scarlet locus. The sex of the flies need not be noted.

<u>Procedure</u>

<u>Day 1 (week 1)</u>: Virgin females of the bw/bw strain are
crossed with st/st males. Each student starts two cultures
with two pairs of flies each.
<u>Day 4 (week 2)</u>: The parental flies are discarded.
<u>Day 12 (week 3)</u>: Anesthetize the progeny and check the eye
color of the flies under the stereomicroscope. Does it cor-
respond to the expected eye color? Start the F_2 by selecting randomly 4 females and 4 males and set up two cultures
with two pairs each.
<u>Day 15 (week 4)</u>: Discard the flies from the vials.
<u>Day 23 (week 5)</u>: Classify the F_2 according to eye color.
Check whether all classes are present approximately in the
expected proportions; exact counts are not required. Relate
each of the different eye colors present to the genotypes
predicted in the checkerboard.

P: Genotypes: ♀ bw/bw; st^+/st^+ x   ♂ bw^+/bw^+; st/st

     Eye color: ....

F_1:

<table>
<tr><td>♂   ♀</td><td></td></tr>
<tr><td></td><td></td></tr>
</table>

Genotypes: ♀   x   ♂ ....

Eye color: ....

F_2:

<table>
<tr><td>♂   ♀</td><td></td><td></td><td></td><td></td></tr>
<tr><td></td><td></td><td></td><td></td><td></td></tr>
<tr><td></td><td></td><td></td><td></td><td></td></tr>
<tr><td></td><td></td><td></td><td></td><td></td></tr>
<tr><td></td><td></td><td></td><td></td><td></td></tr>
</table>

Literature

DICKINSON, W.J., SULLIVAN, D.T.:   Gene-Enzyme Systems in Dro-
    sophila.   In: Results and Problems in Cell Differentia-
    tion, Vol. 6 (eds. W. Beermann et al.).   Berlin, Heidel-
    berg, New York: Springer 1975.

4.4 Chromatographic Analysis of Eye Color Mutants

Specific metabolic mutants are phenotypically recognizable by
their lack of the end product of a specific synthetic pathway. If the missing product is essential for survival, the
defect is lethal. In auxotrophic mutants such as are used in
microbial studies, it is possible to compensate for the defect by supplying the lacking end product in the culture medium. In this way these metabolic mutants can be subcultured
without problem. This, however, is not possible in Drosophila because embryonic and pupal development take place without
the uptake of any food from the outside. Fortunately, there
are several more peripheral, nonessential metabolic pathways
in which mutants can be studied without difficulty. Among
these are the eye color mutants which have only minor effects
on the viability of the carriers of the mutants and thus can
be maintained in homozygous strains. Using such eye color
mutants we can study the basic consequences of a change in a
specific enzyme in a biochemical pathway.

Problem

Determine the various UV fluorescent pterines in the eyes of
the wild type and of some eye color mutants of Drosophila.
Try to develop a hypothesis for the position of the genetic
block present in the biochemical pathways in each mutant.

Material

a) Fly strains: Each student needs 4 adult females (3 to 4
days old) of each of the following stocks: wild type; se/se
(sepia); ry^2/ry^2 (rosy); w^a/w^a (white-apricot); w/w
(white).
b) Filter paper as a working surface, razor blade, watch-
maker's forceps, five short glass rods.
c) Chromatography paper: Each student needs one sheet of
chromatography paper Whatman Nr. 1 which is cut to the right
size to fit into a glass jar of appropriate size (see Figure
16). The size of the sheets should be approximately 23 x 28
cm. For the ascending chromatography applied here, a line is

drawn about 2 cm from the lower end of the paper using a pencil. Five points are marked on this line with an equal distance between them. The heads of the flies will be applied
at these points. Note: Use a pencil. Do not use ball-
point, felt pen or fountain pen as the ink colors would be
chromatographed as well.
d) Chromatography jars: Use a wide-mouthed glass jar of
appropriate size which can be closed airtight (e.g. with a
screw top or with a glass disc which is sealed with Vaseline).
e) Development mixture: Combine 70 ml n-propanol (technical
grade) with 1 ml concentrated ammonia and fill up to 100 ml
with distilled water. The bottom of the jar should be covered with about 1 cm of this mixture.
f) UV lamp: For analyzing the chromatograms a UV lamp is
required. A Philips Hg lamp, model HPW 125 W, fitted in a
normal table lamp is satisfactory if properly shielded to
protect eyes and skin from exposure. See warning later.

<u>Procedure</u>

The wild type females are killed with ether and then placed
on a small piece of filter paper. With the razor blade the
heads of 4 flies are cut off and transferred to the left-most
point on the chromatography paper with the help of a fine
brush or forceps. The 4 heads placed as closely together as
possible are then crushed on the chromatography paper with a
glass rod. This point is labeled + (wild type) using a pen-

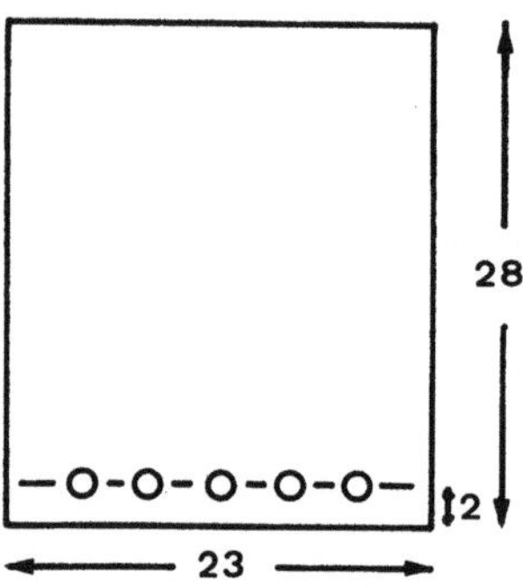

<u>Figure 16</u>. Paper for chromatography. All measurements in cm.

cil. Be careful not to touch the chromatography paper too
often with your fingers as greasy stains will fluoresce yellow later on. The used glass rod is put aside for washing.
Subsequently the heads of the mutants are then applied in the
same way to the remaining points on the chromatography paper. It is important to use a fresh, clean glass rod for
each new mutant so that no eye pigments are transferred from
one point to the next. The chromatogram prepared in this way
is now introduced into the glass jar with the paper curved
into a cylinder held by two paper clips. The chromatogram
should not touch the walls of the glass jar, and the pigment
spots on the paper should not be immersed in the liquid.
Close the glass jar tightly and place it in a draft-free
place. The chromatograms are left to run for 7 to 8 hours.
Afterwards the chromatograms are removed from the tanks and
air-dried by letting them stand on a piece of filter paper
for 2 to 3 hours. If the chromatograms are not analyzed immediately, they should be kept flat in the dark because the
pterines are light-sensitive.

<u>Analysis</u>

View the chromatogram in a dark room under the UV lamp. Mark
the spots that fluoresce with a pencil. The more abundant a
fluorescent compound is, the more intensive is the fluores-
cence.

WARNING:
UV light can damage your eyes! Do not look directly
into the UV lamp in order to avoid eye damage! Glasses
with glass lenses should be worn. Care should also be
taken to avoid excessive exposure of the skin to UV.

In your protocol (see Figure 17) for each mutant record which
compounds are present in normal quantity (+), and which in
increased quantity compared with the wild type (+++ or ++),
and which ones are missing (-).

120

The scheme in Figure 18 gives an overview of the various metabolic pathways involved in the production of the eye pigments. In which step are these pathways blocked in the different mutants? To answer this question, hypotheses may be
developed based on the following two assumptions:
(a) In a given eye color mutant the pigment precursors are
accumulated before a genetic block, whereas substances after
the block are lacking or only present in greatly reduced
quantity.
(b) If a mutant does not show any pigment precursors in this
test, the genetic block may occur in a very early step of the
biosynthetic pathway, or the macromolecular carriers normally
present for the permanent storage of the final pigments in
the eye may be lacking.
Formulate a hypothesis for each of the different mutants.

Literature

BUTENANDT, A., SCHAEFER, W.: Ommochromes. In: Recent Pro-
    gress in the Chemistry of Natural and Synthetic Color-
    ing Matters and Related Fields (eds. T.S. Gore, B.S.
    Joshi, S.V. Sunthankar and B.D. Tilak). New York: Aca-
    demic Press 1962.
HADORN, E., MITCHELL, H.K.: Properties of mutants of _Droso-
    phila_ _melanogaster_ and changes during development as re-
    vealed by paper chromatography. Proc. Nat. Acad. Sci.
    (USA) _37_, 650-665 (1951).
HADORN, E.: Fractionating the fruit fly. Sci. American _206_,
    101-110 (1962).
LANGE, P., WÖHRMANN, K.: Genetisches Grundpraktikum. Stutt-
    gart: Gustav Fischer 1979.
PHILLIPPS, J.P., FORREST, H.S.: Ommochromes and Pteridines.
    In: The Genetics and Biology of Drosophila, Vol. 2d
    (eds. M. Ashburner and T.R.F. Wright). New York: Aca-
    demic Press 1980.
THEOBALD, N., PFLEIDERER, G.W.: Ein neuer Strukturvorschlag
    für die Augenpigmente Drosopterin und Isodrosopterin aus
    Drosophila _melanogaster_. Tetrahydron Lett. _10_, 841-844
    (1977). (Contains structural formulas.)

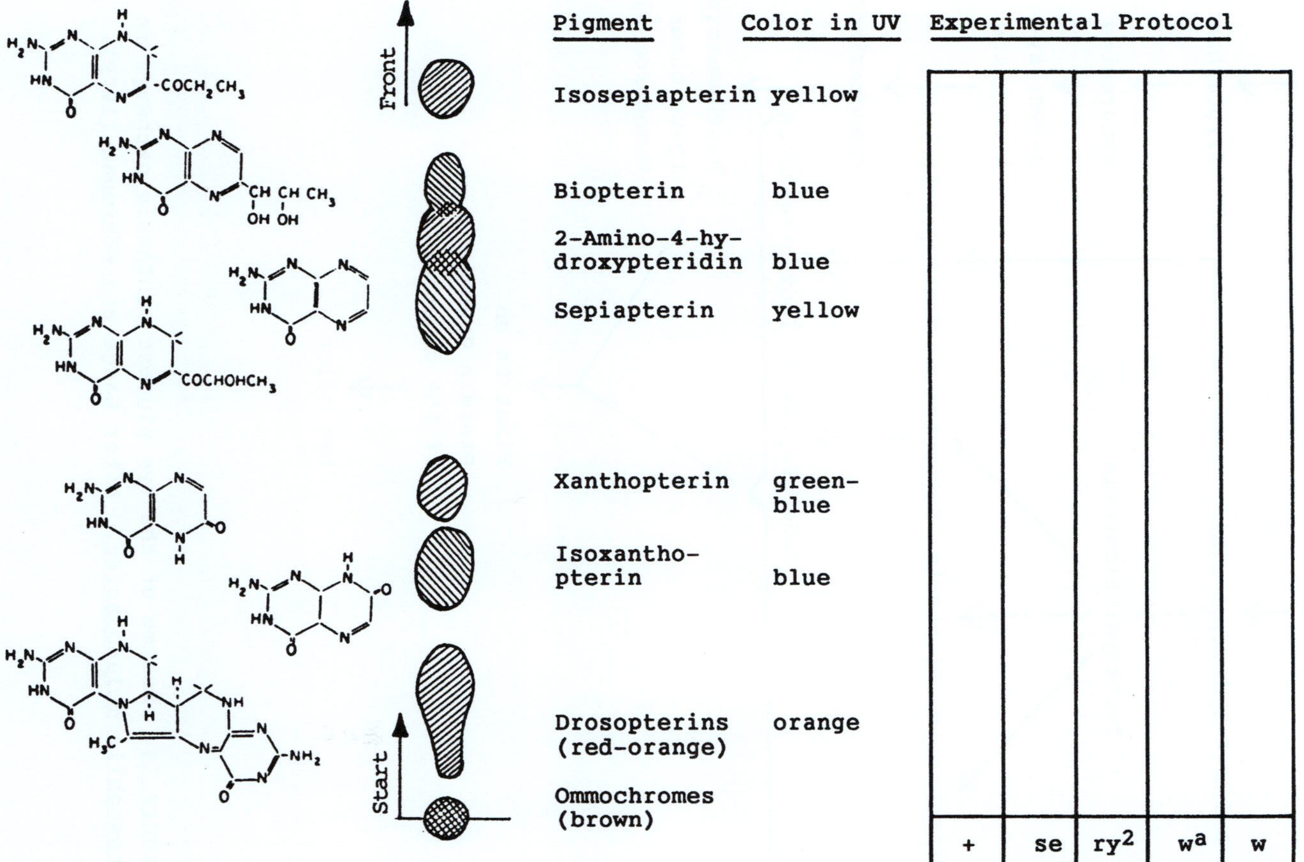

Figure 17. Structural formulas and chromatographic properties of some fluorescing eye pigments in _Drosophila melano-
gaster_.

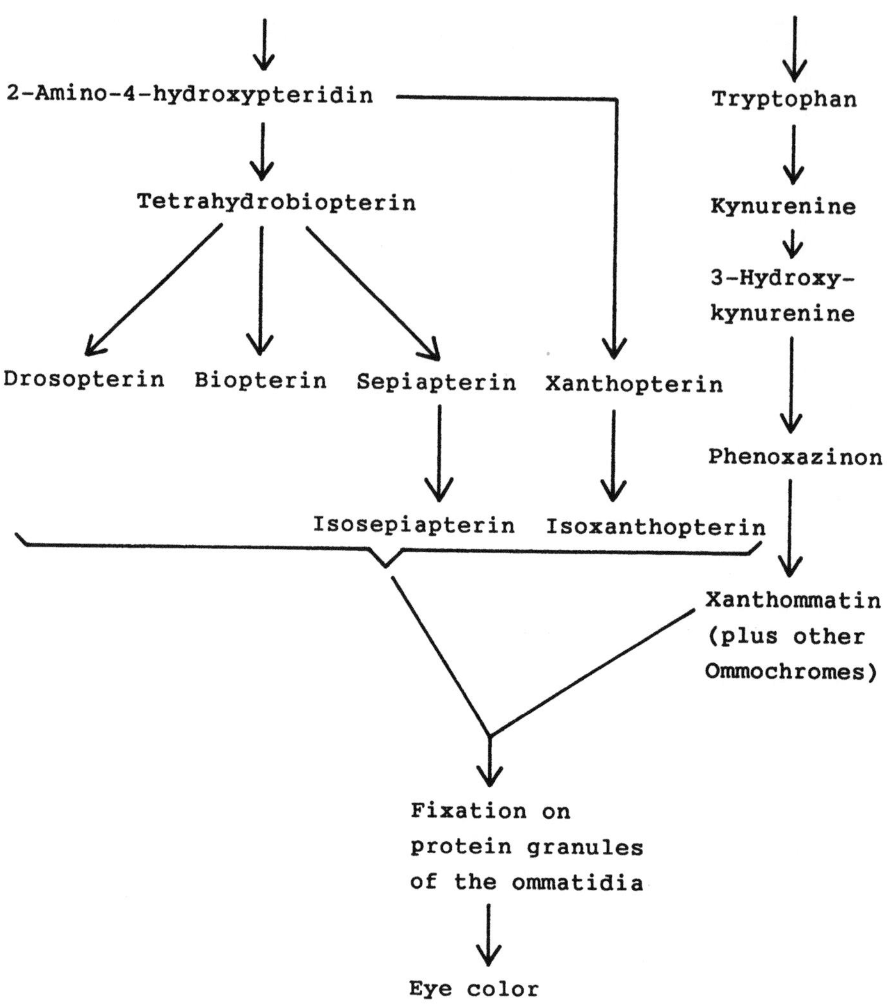

Figure 18. Scheme of the eye pigment synthesis pathways in *Drosophila melanogaster*. After Lange and Wöhrmann (1979).

4.5 Transplantation of Eye Imaginal Discs

The previous experiment has shown that the normal brick red
eye color of the wild type is produced by the presence of
brown and red pigments. We have seen that there are mutants
with brown eye color (due to the lack of the bright red pigments, pterines) as well as others with bright red eye color
(lack of the brown pigments, ommochromes). It has to be assumed that both types of pigments are produced in biosynthetic pathways involving several steps. Therefore it is conceivable that different mutations which block the synthesis
of the ommochromes (leading to a bright red eye color) can
block the biosynthesis in different steps. There are cases
where it is possible to test whether one gene-enzyme system
is able to complement another one which is blocked in a nonallelic step. The following experiment tests whether in the
body of a larva of a given genotype (HOST) the defect present
in a transplanted eye imaginal disc of another genotype (DO-
NOR) can be overcome. If the block in the pathway of the
donor is in a step preceding the block in the pathway of the
host, then the transplanted eye will be able to produce the
normal eye color. In this case the host can help the donor
in getting over its own block by supplying a product needed
after this block. This can be represented schematically as
follows:

Mutant I: HOST $A \xrightarrow{E_1} B \xrightarrow{E_2} C \xrightarrow{e_3} \!\!/\!\!/$ no pigment
(enzyme 3 inactive)

cross feeding

Mutant II: DONOR $A \xrightarrow{E_1} B \xrightarrow{e_2} \!\!/\!\!/ C \xrightarrow{E_3}$ pigment
(enzyme 2 inactive)

The product C synthesized by the host "feeds" the eye imaginal disc of the donor and thus circumvents the block in this
mutant. If the block in the pathway of the host is present
in an earlier step than in the donor, the product required
cannot be supplied and the block is not bypassed.

124

Note: The experiment described below is technically difficult to perform with a whole class. However, it is quite
instructive for the understanding of gene-enzyme systems, and
therefore it has been included here as a mock experiment.
The interpretation of the experimental results is given in
the Results and Answers section.

Problem

Determine the sequence of the defects in the biosynthetic
pathway of the brown eye pigments present in some nonallelic
mutants with the common phenotype bright red eyes.

Material

We suppose that larvae are available from 4 strains. All 3
mutants have bright red eyes:
1)   +/+   wild type
2)   v/v   vermilion   (v, 1 - 33.0)
3)   cn/cn   cinnabar   (cn, 2 - 57.5)
4)   st/st   scarlet   (st, 3 - 44.0)

Procedure

The donor larva is opened and one eye imaginal disc removed.
This imaginal disc is taken up into a micropipette and then
injected into the body cavity of an anesthetized host larva.
The host larva with the implanted donor disc is raised on
standard medium. After pupation a host fly will eclose. In
the abdomen of the host fly the implant can be found which
has metamorphosed into an eye during pupal development. The
eye color of this implant can be recorded (see Figure 19).

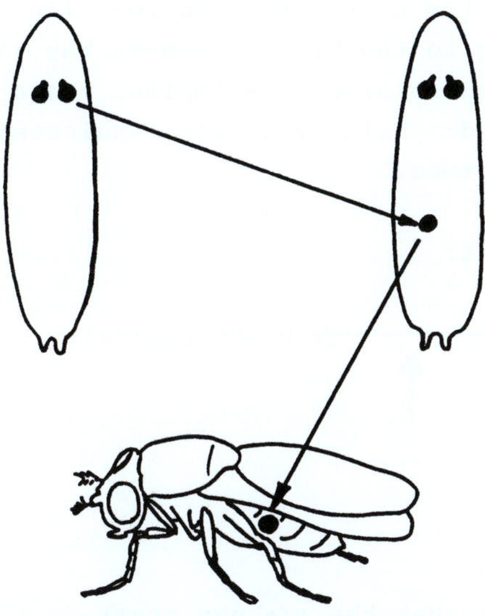

<u>Figure 19</u>. Transplantation of an eye imaginal disc. After Wagner and Mitchell (1955).

<u>Results</u>

Genotype of host	Genotype of donor	Color of the donor eye in the host fly
+/+	+/+	brick red
+/+	v/v	brick red
+/+	cn/cn	brick red
v/v	+/+	brick red
v/v	v/v	bright red
v/v	cn/cn	bright red
v/v	st/st	bright red
st/st	+/+	brick red
st/st	v/v	brick red
st/st	cn/cn	brick red
st/st	st/st	bright red
cn/cn	+/+	brick red
cn/cn	v/v	brick red
cn/cn	cn/cn	bright red
cn/cn	st/st	bright red

As a first step the sequence of the different mutations (cn, st and v) in the pathway leading to the brown pigments has to be determined. Note: Bright red color of the implant means that no brown pigments are formed; brick red color indicates that brown pigments have been formed.

The following sequence is present:

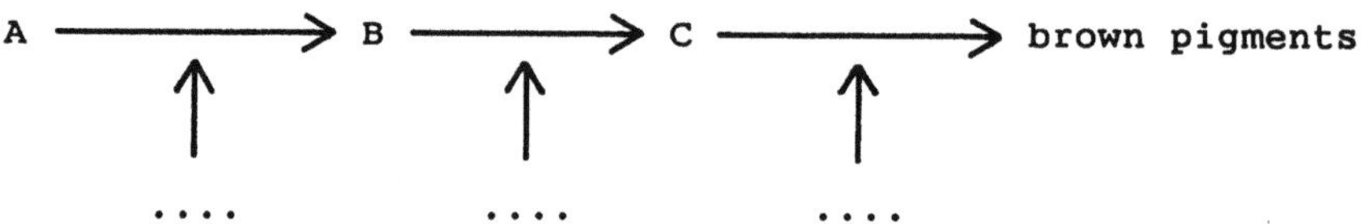

Two additional problems to solve:
(a) How do you explain the fact that the +/+ eye transplanted into any host develops a normal brick red eye color?
(b) In the table with the results on the previous page the following experiment has been left out: In a +/+ host an st/st implant shows a bright red eye color and NOT a brick red one as expected. Does this mean that the sequence determined as above is not correct? Are there other possibilities to explain this unexpected experimental result?

Literature

WAGNER, R.P., MITCHELL, H.K.: Genetics and Metabolism. New York: John Wiley and Sons 1955.

4.6 Supplementation of an Eye Color Mutant

In the previous experiment "Transplantation of eye imaginal discs" we have seen that in the three mutants vermilion, cinnabar and scarlet, the synthesis pathway of the brown eye pigments (ommochromes) is blocked in three different positions. There is another way to demonstrate the existence of such genetic blocks; by adding an intermediate product lacking in a specific mutant it is possible to abolish the block phenotypically (supplementation).

Problem

The mutant vermilion is not able to convert tryptophan into kynurenine due to the lack of tryptophan oxygenase. By supplementing the food of larvae of this strain with kynurenine we shall attempt to produce a normal eye color in the resulting adults (phenocopy). From the order of gene action worked out in experiment 4.5 how would you expect the mutant cinnabar to react to feeding of larvae with kynurenine?

Material

Sometimes it is difficult to distinguish the vermilion eye color from the normal eye color. For this reason we use double mutants: The strain v; bw as well as the strain cn bw has white eyes because in both the synthesis of the red pigments (bw) as well as of the brown pigments (v and cn, respectively) is blocked (see also experiments 4.3 and 4.5). Adult females and males of both strains are needed. In addition the following two solutions are prepared:
Antibiotics (A):   0.05% streptomycin plus 0.033% penicillin
                    in distilled water.
Kynurenine: 1% DL-kynurenine in solution A.
To apply the solutions it is best to use a 200 μl pipette or a 2.5 ml graded syringe without needle. Furthermore a glass rod (0.5 cm diameter) is needed.

Experimental work and evaluation

Day 1: Start 2 cultures each of the 2 strains v; bw and cn
bw with 4 pairs of flies.

Day 2: The parents are discarded. From each strain one vial
is treated with the antibiotics solution and one with the
kynurenine solution. The treatment is done in the following
way: With a glass rod make a hole in the medium of a few
millimeters depth and place 200 µl of the solution in the
hole. This solution is now taken up by the larvae developing
in the vial.

Day 4: The treatment of all the vials is repeated again in
the same way.

Day 6: The treatment is repeated once more.

Day 12: The eclosing flies are anesthetized and their eye
color is recorded. Which flies have white eyes? Which flies
have brown eyes?

Results

Strain v; bw + antibiotics :   ....

Strain v; bw + kynurenine :   ....,.

Strain cn bw + antibiotics :   ....

Strain cn bw + kynurenine :   ....

Literature

BUTENANDT, A., WEIDEL, W., BECKER, E.: Kynurenine als Augen-
     pigmentbildung auslösendes Agens bei Insekten. Naturwis-
     senschaften 28, 63-64 (1940).
GREEN, M.M.: A study of tryptophan in eye color mutants of
     Drosophila. Genetics 34, 564-572 (1949).
MUCKENTHALER, F.A.: Kynurenine localization in the egg of
     Drosophila melanogaster. Experientia 27, 828-830 (1971).
PARSONS, P.A., GREEN, M.M.: Pleiotropy and competition at
     the vermilion locus in Drosophila melanogaster. Genetics
     45, 993-996 (1959).
POTTER, J.H.: A demonstration of compensation for an inher-
     ited biochemical defect in Drosophila melanogaster. Dro-
     sophila Inf. Serv. 47, 134 (1971).

4.7 Genetic Complementation and Allelism

<u>Changes in different genes may produce similar phenotypes</u>

As was seen in the previous experiments, biochemical pathways
resulting in an end product such as a particular pigment
consist of a number of different discrete steps each under
the control of a different gene locus. A mutation of any one
of these genes will produce a phenotype showing a change in
the single end product of the pathway. This means that mutations in entirely different genes may produce the same gross
phenotype. For example the mutants v (vermilion) on the X
chromosome, cn (cinnabar) on the 2nd chromosome and st (scarlet) on the 3rd chromosome all produce a bright red phenotype
and are indistinguishable from each other phenotypically.
This must always be kept in mind especially when dealing with
the genetics of organisms less well understood than that of
Drosophila. For example in human genetics, it must always be
remembered that two hereditary conditions, which have the
same symptoms identified in two different families, may be
the same genetically or may be due to changes in different
genes.

<u>Different changes in the same gene may produce different
phenotypes</u>

Since genes are made up of several thousand nucleotide pairs
in a DNA molecule, each normal (wild type) gene could poten-
tially change in many possible ways. Each changed form of
the gene is a different allele. Some alleles may produce no
visible phenotypic effect while others may have a drastic
effect. This means that different changes in the same gene
may produce different phenotypes.

<u>Complementation test (allelism test)</u>

Suppose that you had discovered a new mutation to bright red
eyes in some wild flies. How would you determine if this is
an allele of a known mutant of the same phenotype (e.g. st)
or a change in another gene?

To solve this problem, we can carry out a complementation or allelism test. In this test, the unknown recessive mutant (homozygous) is crossed to the known recessive mutant (also homozygous). If the two are alleles of each other, no normal allele will be present in the heterozygote and its phenotype will be mutant. If, however, the two are not due to changes in the same gene (nonallelic), each will carry the wild type allele of the other mutant and the F_1 will be heterozygous for both mutant and wild type alleles and will accordingly show a wild type phenotype. Mutants in two different genes can complement each other since each carries the normal gene lacking in the other. Two mutant alleles of the same gene cannot complement each other.

<u>Problem</u>

Determine by making appropriate crosses, whether the gene producing a black body color in the females of your A strain is allelic to the genes causing black body color in the males of strains B and C, respectively.

<u>Material</u>

Virgin females of either b/b (black, 2-48.5; black body color) or e/e (ebony, 3-70.7; ebony body color) labeled A (coded by instructor). Males of e/e and se/se (sepia, 3-26.0; sepia eye color) labeled B and C, respectively (coded by instructor).

<u>Procedure</u>

<u>Day 1 (week 1)</u>: Make these two crosses with several pairs of flies per vial:
(1)   ♀ A   x   ♂ B
(2)   ♀ A   x   ♂ C
<u>Day 4 (week 2)</u>: Discard the parents from both crosses.
<u>Day 12 (week 3)</u>: Classify the progeny from each cross for body color.

<u>Results</u>

Cross	Phenotype of F_1	Complementation?
(1) A x B		
(2) A x C		

<u>Conclusions</u>

(a)   Is the mutation present in strain A allelic to that in strain B?

(b)   Is the mutation present in strain A allelic to that in strain C?

<u>Literature</u>

HARTL, D.L., FREIFELDER, D., SNYDER, L.A.: Basic Genetics.
    Boston MA: Jones and Bartlett 1988.
KLUG, W.S., CUMMINGS, M.R.: Concepts of Genetics, 2nd ed.
    Columbus OH: Merrill Publishing Co. 1986.
LEWIN, B.: Gene Expression, vol. 2: Eukaryotic Chromosomes,
    2nd ed. New York: John Wiley and Sons 1980.

4.8 Transposable Elements

Until fairly recently, it was assumed that most mutations in
Drosophila (and in other organisms) represented simple
changes in DNA such as single base pair substitutions or deletions or additions of one or more base pairs. In addition
to these, it was appreciated that chromosomal aberrations
could produce mutational effects because of the disruption of
genes and regions of DNA involved in regulation that occurred
when chromosomes were broken and rearranged. Today it is
known that there is an entirely different type of event that
is responsible for a great deal of genetic alteration, the
insertion of transposable elements or "jumping genes". These
elements, first discovered in maize over forty years ago have
recently been found to be present in every type of organism
where appropriate systems for their detection exist. It came
as a real surprise to most Drosophila geneticists when "old"
spontaneous mutants that had been cultured in laboratories
for many years were analyzed on the molecular level and were
found to be mutant because of the insertion of a transposable
element into a structural gene rather than because of simple
chemical changes. There is no doubt that the activity of
these elements, which can move around the genome under certain conditions, is a very important source of genetic variation in Drosophila. There are different classes of mobile
elements in Drosophila, and one of the best studied and easiest types of element to demonstrate is the class known as the
P elements, which were originally discovered because of the
phenotypic effects on sterility produced when males having P
elements are crossed with females lacking them. The P element has been sequenced and is known to be 2907 base pairs in
length and to be flanked by 31 base pair inverted repeats.
In addition to complete elements, many defective elements exist which have regions of varying size deleted.

Demonstration of P-M hybrid dysgenesis

Strains needed: (1) A P line preferably with visible markers. Such a line has multiple complete P elements which
when active are capable of producing a "transposase" enzyme

which allows transposition. P lines are said to have a P cytotype in which P elements are stable in position.
(2) An M line. This is a line lacking P elements. Most laboratory strains established before 1960 are M strains. M lines have an M cytotype.

<u>P-M hybrid dysgenesis</u>: The easiest effect of P element activity to observe is the temperature sensitive hybrid dysgenesis produced when active P elements are introduced into an M cytotype. This results in the arrestment of development of one or both gonads in the hybrid flies when they are cultured at high temperatures (above 23°C).

<u>Procedure</u>

Each student should set up the following four crosses:

<u>Cross</u>	<u>Female</u>	<u>Male</u>	<u>Culture Temp</u>.
1	w^+(P)	w (M)	20°C
2	w^+(P)	w (M)	28°C
3	w (M)	w^+(P)	20°C
4	w (M)	w^+(P)	28°C

<u>Day 1 (week 1)</u>: Make the four crosses listed above in vials and immediately place the vials in incubators at the indicated temperature. If the laboratory is cool and the temperature does not go above 23°C, the low temperature crosses may be left at room temperature.
<u>Day 4 (week 2)</u>: Discard the parents. Since the gonadal development is arrested during the first four days care should be taken to keep the temperatures constant during this period.
<u>Day 12 (week 3)</u>: Anesthetize and kill the F_1 progeny.
Dissect the abdomens of 20 females from each cross. Place each female for dissection in a drop of water or Ringer solution in a well slide or watch glass over a dark background and tear open the abdominal wall with a forceps or needle while holding the fly by the thorax with forceps. The ovaries can then be examined under the dissecting microscope.
Fertile females will have ovaries packed full of developing eggs while those having dysgenic ovaries will have one or both ovaries which are empty of eggs (see Figure 20).

Record your results for each cross in the following table.

Cross	Number of females with:			Dysgenic ovaries (%)
	Both ovaries normal	One ovary normal	No normal ovaries	
1				
2				
3				
4				

In which cross do you observe dysgenic ovaries? Can you think of any other crosses that might have been done in this experiment?

<u>Figure 20</u>. Drawing of a fertile and a sterile ovary.

<u>Literature</u>

KIDWELL, M.G.: P-M mutagenesis. In: Drosophila, a Practical Approach (ed. D.B. Roberts). Oxford, Washington DC: IRL Press 1986.

5. Mutation Genetics

5.1 Induction and Detection of Sex-Linked Recessive Lethals

Various test systems are known in <u>Drosophila</u> <u>melanogaster</u> for the detection of the different types of mutations. One of the most frequently used mutagenicity test systems detects sex-linked recessive lethals (SLRL). With this test any point mutation or chromosome aberration occurring on the X chromosome which is lethal in hemizygous condition can be detected. The test takes two generations because it is restricted to the detection of the mutations induced in the X chromosome only. One takes advantage of the fact that the recessive lethals manifest themselves in hemizygous (XY) males, but are not expressed in heterozygous females and can be transmitted to the next generation. The "Muller-5" or "Basc" method (see Figure 21) makes use of a so-called "balancer chromosome" which carries two inversions as well as the dominant marker Bar (B, bar eyes) and the recessive marker white-apricot (w^a, apricot eyes).

For the test, wild type males are treated with a mutagenic agent. When using chemical mutagens, the substances are usually fed to the males for one to three days. When using ionizing radiation, the males are irradiated acutely. These treatments induce mutations, and those in the germ cells which are in the various stages of spermatogenesis in the gonads of the males can be detected by this test. For the detection of recessive lethals the males are crossed to homozygous Basc females. Each sperm containing an X chromosome results, after fertilization, in a heterozygous F_1 female which is then tested for the presence of a recessive lethal. Due to the inversion heterozygosity of these F_1 females, crossover products are eliminated, and each female transmits only unaltered copies of the X chromosome to be tested, to the F_2 progeny (see also experiment 7.2). For this, each F_1 female is placed in a vial with one F_1 male which is hemizygous for the Basc chromosome. The females need not be virgins since their brothers have the desired genotype. In the F_2, among other classes of progeny, males are produced that are hemizygous for the X chromosome to be tested. If the X chromosome carries a recessive lethal, this specific class of males will be missing in the respective F_2 culture.

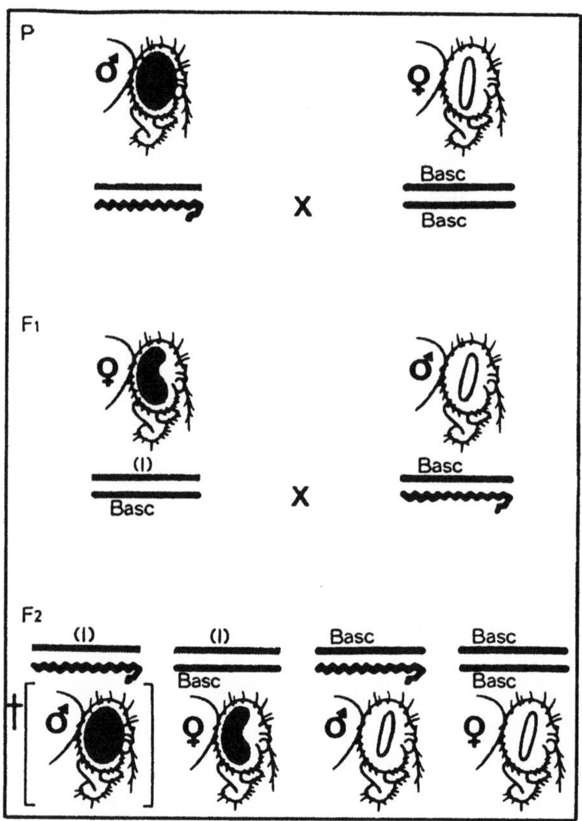

Figure 21. Schematic representation of the Basc test for the detection of sex-linked recessive lethals. X chromosomes treated in the P males are tested for the presence of a recessive lethal (l). When such a mutation has been induced, the wild type males with normal red eyes are missing in the F_2 (lower left in the diagram). After Würgler et al. (1983).

The mutation rate can then be calculated as follows:

$$\text{SLRL (\%)} = \frac{\text{number of cultures with lethals}}{\text{total number of cultures}} \times 100$$

The induced mutation rate is calculated as:

Induced rate = rate in treated series - rate in control

<u>Problem</u>

Determine the frequency of sex-linked recessive lethals in
mature sperm of males which have been fed for 24 h with a
solution of the alkylating agent EMS (ethyl methanesulfonate)
and calculate the rate of induced mutations. A range of con-
centrations between 0.1 mM and 25 mM EMS can be used. (Note:
X-irradiation can also be used; suggested dosage 20 Gy).

<u>Material</u>

Adult wild type males and virgin females of the Basc strain
[Basc: In(1)sc^{S1L}sc^{8R}+S, sc^{S1}sc^{8} w^{a} B]. These females have
apricot Bar eyes.

<u>Preparation of chemical mutagen solutions</u>

WARNING
EMS is mutagenic and carcinogenic in
experiments with mammals (IARC, 1974).
Follow safety precautions strictly!

<u>Safety precautions</u>: For the preparation of the mutagen solu-
tions the following measures should be taken: 1. Work in a
ventilated hood. 2. Cover working surface with appropriate
protective paper (or aluminum foil). 3. Wear rubber gloves.
4. Use disposable material if possible, which can be collected for incineration in a closed plastic bag. 5. Use
automatic pipettes with disposable tips. No mouth pipetting!
6. Unused mutagen solutions and glassware can be decontami-
nated with an 0.5% solution of thioglycolic acid (TGA) in 1 N
potassium or sodium hydroxide solution (e.g. 28 g KOH in 500
ml H$_2$O plus 2.5 ml TGA). Fill glassware with this solution
and let stand for a few days.

<u>Phosphate buffer</u>: 5.9 g Na$_2$HPO$_4$.2H$_2$O per liter; 4.5 g
KH$_2$PO$_4$ per liter; mixture in a 1 : 1 ratio gives pH 6.8.
<u>EMS solution</u>: 200 ml phosphate buffer plus 10 g sucrose plus
0.48 ml ethyl methanesulfonate (EMS, M$_r$ 124) gives a 25 mM
EMS feeding solution.

<u>Stability of the solution</u>: Aqueous solutions of the alkylating agent EMS are relatively unstable at room temperature (half life 11.6 h). Use of phosphate buffer makes the solutions more stable; however solutions should always be made up freshly for immediate use.

<u>Feeding</u>: A tissue is crumpled into an empty Drosophila culture bottle and pressed firmly to the bottom. 9 to 10 ml of mutagen solution are then added with a pipette so that the tissue is just soaked but no free liquid present. Subsequently 50 to 100 adult males are introduced into the feeding bottle which is then closed with a cotton or foam rubber stopper. The males to be treated (2 to 3 days old and well fed) are first starved for 4 to 5 h in an empty culture bottle. The males should be transferred into the feeding bottle without anesthesia so that they do not stick to the tissue. The males are normally fed for 24 h. At the end of the feeding period they are transferred into a fresh culture bottle with standard medium and can then be used for crosses to appropriate virgin females. A control series is treated in the same way but fed only 5% sucrose solution.

<u>Experiment</u>

Complete the following two checkerboards.

P:　♀ Basc, w^a B / Basc, w^a B　x　♂　X / Y

(EMS treated)

Phenotype of F_1:

....

....

Enter genotype and phenotype (sex, eye color, eye shape) of the F_1 and F_2 flies. Use F_1 flies to start the following cross:

F_1:　♀　Basc, w^a B / X (EMS)　　x　　♂　Basc, w^a B / Y

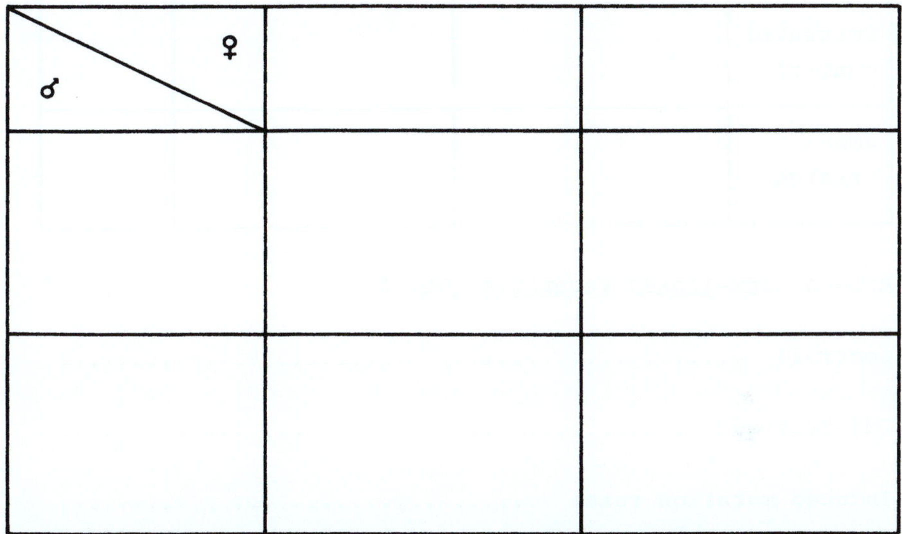

Which F_2 class is missing when a recessive lethal has been induced in the treated X chromosome? Why is it not necessary to discard the parents (F_1 flies) from the cultures? Why do the F_1 females have to be crossed individually?

Procedure

<u>Day 1 (week 1)</u>: The wild type males are fed with 25 mM EMS for 24 h. For the negative control males are fed sucrose solution. One treated male is crossed with 3 Basc females per vial.

<u>Day 4 (week 2)</u>: The parents are discarded.

<u>Day 12 (week 3)</u>: Each F_1 female is placed in a separate vial with an F_1 male. At least 200 crosses each should be set up from the treated and the control series.

<u>Day 23-26 (week 5)</u>: The F_2 flies of each individual vial are anesthetized and checked for the presence of wild type males (round red eyes). Absence of the normal males indicates a recessive lethal. Only vials with at least 10 to 15 progeny are included.

Results

	Vials without lethal (normal cultures)	Vials with lethal (cultures without wild type ♂)	Total
Untreated control			
EMS-treated			

Rate of sex-linked recessive lethals

Control: ....

EMS-treated: ....

Induced mutation rate: ....

Literature

IARC: Some anti-thyroid and related substances, nitrofurans
    and industrial chemicals. IARC Monographs on the evalua-
    tion of carcinogenic risk of chemicals to man, vol. 7.
    Lyon: International Agency for Research on Cancer 1974.
LEWIS, E.B., BACHER, F.: Method of feeding ethyl methanesul-
    fonate (EMS) to Drosophila males. Drosophila Inf. Serv.
    43, 193 (1968).
MULLER, H.J.: Artificial transmutation of the gene. Science
    66, 84-87 (1927). In: Classic Papers in Genetics (ed.
    J.A. Peters). Englewood Cliffs NJ: Prentice-Hall 1959.
VOGEL, E., LEIGH, B.: Concentration-effect studies with MMS,
    TEB, 2,4,6-triCl-PDMT, and DEN on the induction of domi-
    nant and recessive lethals, chromosome loss and trans-
    locations in Drosophila sperm. Mutation Res. 29, 383-396
    (1975).
VOGEL, E., SCHALET, A., LEE, W.R., WüRGLER, F.E.: Mutagenic-
    ity of selected chemicals in Drosophila. In: Compara-
    tive Chemical Mutagenesis (eds. F.J. de Serres and M.D.
    Shelby). New York, London: Plenum Press 1981.
WüRGLER, F.E., SOBELS, F.H., VOGEL, E.: Drosophila as assay
    system for detecting genetic changes. In: Handbook of
    Mutagenicity Test Procedures, 2nd ed. (eds. B.J. Kilbey
    et al.). Amsterdam: Elsevier/North Holland Biomedical
    Press 1983.

5.2 Induction and Detection of Reciprocal Translocations

In the testing of compounds for genotoxic effects, it is desirable to be able to identify those that produce breaks in chromosomes (clastogens). Since one of the results of chromosome breaks is the production of chromosome aberrations, a test that detects the production of a specific type of aberration, in this case reciprocal translocations, can give an unambiguous test of this type of chromosome damage. The test takes two generations. Specific types of reciprocal translocations are detected by the absence of specific classes of progeny in the F_2. The test starts with a cross of mutagen-treated wild type males with females carrying eye color markers on chromosomes 2 and 3. From the resulting F_1 flies only the males are used for further testing because crossing over in the females interferes with the demonstration of translocations. The F_1 males are test crossed individually to genetically marked females. Due to the segregation of the autosomal markers, eight different classes of progeny are expected in the F_2. When a translocation changing the linkage group of one or more markers has been induced in the mature sperm in the treated P males, four classes of progeny will be missing from the F_2, because the segregation of the translocation chromosomes produces aneuploid nonviable genotypes. Depending on the type of translocation induced (between the autosomes 2 and 3, between Y and 2, or between Y and 3), specific classes of progeny will be missing.

Problem

Determine the rate of reciprocal translocations after irradiation of wild type males with 40 Gy X-rays. (X-rays produce much higher rates of translocations than an EMS treatment).

Material

Adult wild type males. Virgin females with three recessive autosomal markers: bw/bw; st p^p/st p^p. The marker brown (bw, 2-104.5) alone produces a brown eye color, scarlet (st,

3-44.0) alone results in light red eyes, and pink-peach
(p^P, 3-48.0) alone gives a peach eye color. Flies of the
triply marked homozygous strain (bw/bw; st p^P/st p^P) have
white eyes.

Experiment

Simplified symbols are used to fill in the following tables.
The wild type males possess unmarked chromosomes: X/Y; 2/2;
3/3, the white-eyed females have marked autosomes (bw; st
p^P): X/X; 2*/2*; 3*/3*.

P : ♀ X/X; 2*/2*; 3*/3* x ♂ X/Y; 2/2; 3/3 (irradiated)

♀ \ ♂		

What is the phenotype of the resulting F_1 flies? To avoid
the complications produced by crossing over in the heterozygous F_1 females, only the males are used for further crossing. Each F_1 male represents <u>one</u> irradiated sperm. For
the test cross we use again triply marked homozygous females.

F_1 : ♀ X/X; 2*/2*; 3*/3*   x   ♂ X/Y; 2/2*; 3/3*

From this cross the following eight classes of progeny are
expected (use the simplified symbols):

Class:	A	B	C	D	E	F	G	H
♀ \ ♂	X23	Y23	X2*3	Y2*3	X23*	Y23*	X2*3*	Y2*3*
X2*3*								
Sex								
Eye color								

```
2*/2*  =  bw/bw  =  brown eyes
3*/3*  =  st pᴾ/st pᴾ  =  light yellow eyes
2*/2*; 3*/3*  =  bw/bw; st pᴾ/st pᴾ  =  white eyes
```

$2^*/2^* = bw/bw =$ brown eyes

$3^*/3^* = st\ p^P/st\ p^P =$ light yellow eyes

$2^*/2^*;\ 3^*/3^* = bw/bw;\ st\ p^P/st\ p^P =$ white eyes

All these eight classes are produced when <u>no</u> translocation
has been induced. Some of the sperm of the P males may carry
translocations. The three types of reciprocal translocations
that can be distinguished are shown below. The circles represent centromeres. The marker genes are represented by a
single vertical mark for bw and a double mark for st p^P.

Chromosomal constitution	Symbol	Type
normal: 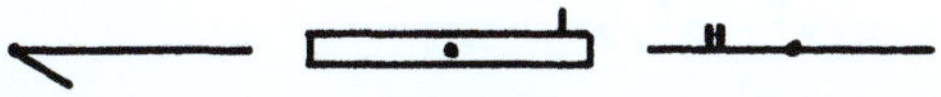	$Y\ 2\ 3$	normal
Translocation between 2 and 3: 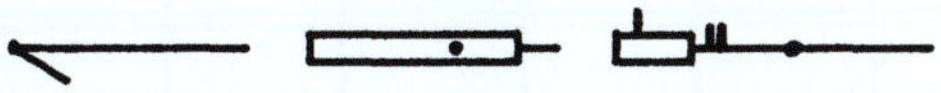	$Y\ 2^T\ 3^T$	T1
Translocation between Y and 2: 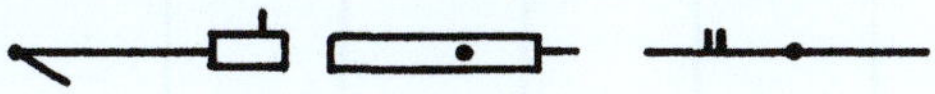	$Y^T\ 2^T\ 3$	T2
Translocation between Y and 3: 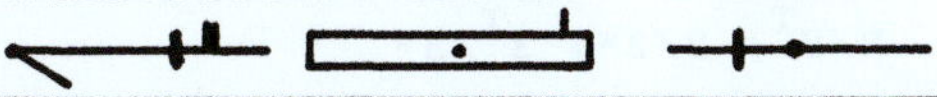	$Y^T\ 2\ 3^T$	T3

The example of a translocation of type T1 will show the consequences. In this case the F_1 male has the chromosomal
constitution X/Y; $2^T/2^*$; $3^T/3^*$. It also produces eight
classes of gametes just as a normal male would do (see F_1
crossing scheme). As examples look at two classes of gametes:

Class B

♀ ╲ ♂	$Y\ 2^T 3^T$
X 2*3*	

Class D

♀ ╲ ♂	$Y\ 2^* 3^T$
X 2*3*	

Would you expect these classes of progeny to be viable or
not? Why?

Work out the possible combinations expected in the test cross progeny for the three types of translocations using euploid as well as aneuploid gametes:

Translocation T1: $F_1 - \male$: X/Y; $2^T/2*$; $3^T/3*$

Class	A	B	C	D	E	F	G	H
♀＼♂								
X 2* 3*								
Phenotype								

Translocation T2: $F_1 - \male$: X/Y^T; $2^T/2*$; $3/3*$

Class	A	B	C	D	E	F	G	H
♀＼♂								
X 2* 3*								
Phenotype								

Translocation T3: $F_1 - \male$: X/Y^T; $2/2*$; $3^T/3*$

Class	A	B	C	D	E	F	G	H
♀＼♂								
X 2* 3*								
Phenotype								

Indicate the nonviable classes by putting an **X** in the relevant square. Determine sex and eye color of the viable classes.

Procedure

__Day 1__: At least 60 wild type males are irradiated with 40 Gy X-rays (45 keV, 10 mA, dose rate 20 Gy/min) and crossed with virgin bw; st p^p females. One male and three females per vial are used. For the negative control about 20 unirradiated males are used.

__Day 4__: The parental flies are discarded from the vials.

__Day 12__: The F_1 males are testcrossed __individually__ with virgin bw; st p^p females; two to three females per vial are added. In the control series about 100 vials and in the treated series at least 200 vials are set up.

__Days 23 to 26__: The progeny in the vials are anesthetized separately. In each vial the presence of all the eight phenotypic classes (females and males, four different eye colors) has to be verified (= normal). The absence of specific classes of flies indicates a translocation of a specific type (= T1, T2 or T3). Only vials with at least 10 progeny are included.

Results

	Number of vials				Total
	Normal	T1	T2	T3	
Unirradiated control					
Irradiated					

Rate of reciprocal translocations

Control   : ....

Irradiated: ....

Literature

GONZALES, F.W.:   Dose response kinetics of genetic effects induced by 250 kVp X-rays and 0.68 Mev neutrons in mature sperm of __Drosophila__ __melanogaster__. Mutation Res. __15__, 303-324 (1972).

5.3 Mutagen-Induced Sex Chromosome Losses

One of the aims of genetic toxicology is the detection of
chromosome damaging activity of chemical substances in test
systems that are as fast and easy to perform as possible.
One very drastic consequence of chromosome damage is the loss
of complete chromosomes. This event is easily detectable in
<u>Drosophila</u> <u>melanogaster</u> with its simple chromosome complement
(n=4). The loss of one of the two large autosomes leads to
dominant lethality due to dosage effects of many genes. In
contrast to this, some individuals with the loss of a sex
chromosome can be viable. When an X chromosome is lost from
an X/X zygote or when the Y chromosome is lost from an X/Y
zygote, viable X/0 males result in both cases. This phenomenon can be applied for a simple mutagenicity test that takes
only one generation. Using suitably marked chromosomes, sex
chromosome losses induced in mature sperm by ionizing radiation or chemical mutagens can be detected. To increase the
efficiency of the test further, one can employ a second phenomenon: Ring-shaped X chromosomes are known in Drosophila.
These ring-X chromosomes have the characteristic that they
are lost much more frequently than normal rod-X chromosomes
both spontaneously and after mutagen treatment. The mechanisms leading to the more frequent loss of ring-X chromosomes
are not yet fully understood: Ring chromosome losses can be
either the consequence of chromosome breakage events or of
sister chromatid exchanges (SCE).

The rates of sex chromosome losses registered experimentally
can be influenced further by various other factors: Mutagen-
treated male germ cells can be tested for chromosome losses
by crossing the treated males with females of different genetic constitution. The resulting rates of sex chromosome
losses can differ significantly, e.g. there are strong maternal effects on the chromosome losses in the paternal genome.
Such maternal effects are most pronounced when females are
used which are homozygous for the meiotic mutation mei-9[a].
This mutant is defective in cellular DNA excision repair.

Problem

Males carrying both a marked ring-X as well as a marked Y
chromosome are treated with a chemical mutagen or irradiated
with X-rays and then crossed to excision repair competent and
to excision repair defective (mei-9) females which have not
been treated. To determine the frequency of sex chromosome
losses induced in mature sperm of the ring-X males, we use a
ring-X chromosome [X^{c2}, new designation R(1)2] which carries the markers y (yellow body color) and f (forked bristles); the Y chromosome is marked with y^+ (normal body col-
or) and B^S (Bar of Stone, bar eyes). Besides complete
losses of the whole X or Y chromosome, partial losses of the
Y chromosome are also possible. This can be detected by the
loss of only the one or the other of the two markers (y^+ or
B^S, respectively) on the Y chromosome.

Material

Flies used: virgin females: y w
$$y\ w^a\ mei\text{-}9^a$$
males: $X^{c2}, y\ f\ /\ y^+\ Y\ B^S$

Experiment

On the next two pages checkerboards are prepared for the two
different crosses. When completing these remember that the
loss of the X or Y or parts of the Y chromosome in mature
sperm of the males leads to additional types of male gametes. Furthermore, in the cross with the meiotic mutant
nondisjunction of the sex chromosomes in the females has to
be taken into account as well (see experiment 3.8).

Procedure

Day 1: Collect virgin females of the two strains y w and y
w^a mei-9^a.
Day 2: Collect males of the strain X^{c2}. If necessary,
collect more virgin females.
Day 3: Starve the males for 4 to 5 h in an empty culture

<u>Cross 1:</u>   Gametes, genotypes and phenotypes

♀ y w   x   ♂ X^{c2}, y f / y^+ Y B^S

Enter the phenotypes: Sex, eye shape, eye color, body color and lethality if applicable.

<table>
<tr><td rowspan="2">♂ ╲ ♀</td><td>normal</td><td colspan="2">nondisjunction</td></tr>
<tr><td></td><td></td><td></td></tr>
<tr><td rowspan="2">normal</td><td></td><td></td><td></td></tr>
<tr><td></td><td></td><td></td></tr>
<tr><td>losses</td><td></td><td></td><td></td></tr>
<tr><td rowspan="2">Y partial losses (PL)</td><td></td><td></td><td></td></tr>
<tr><td></td><td></td><td></td></tr>
</table>

<u>Cross 2</u>: Gametes, genotypes and phenotypes

$\female$ y w^a mei-9^a x $\male$ X^{C2}, y f / y$^+$ Y B^S

Enter the phenotypes: Sex, eye shape, eye color, body color and lethality if applicable.

<table>
<tr><td rowspan="2"></td><td>normal</td><td colspan="2">nondisjunction</td></tr>
<tr><td>$\female$
$\male$</td><td></td><td></td></tr>
<tr><td rowspan="2">normal</td><td></td><td></td><td></td></tr>
<tr><td></td><td></td><td></td></tr>
<tr><td>losses</td><td></td><td></td><td></td></tr>
<tr><td>Y partial losses (PL)</td><td></td><td></td><td></td></tr>
</table>

bottle. Feed the males with an 0.6 mM MMS (methyl methanesulfonate) solution (for method see experiment 5.1).

<u>MMS solution</u>: 200 ml phosphate buffer plus 10 g sucrose plus
10 µl methyl methanesulfonate (MMS, M_r 110) gives an 0.6
mM MMS feeding solution. See experiment 5.1 for safety precautions!

Instead of MMS, 20 Gy X-rays (45 keV, 10 mA, dose rate 20
Gy/min) can also be used as a mutagenic agent.

<u>Day 4</u>: End feeding after 24 h. Cross males with females in
vials as follows:

Series y w :   one pair per vial
Series mei-9[a]:   3-5 pairs per vial

Use untreated males for the negative controls.

<u>Day 7</u>: Discard parental flies from the vials. The insemi-
nated females can be used again for establishing a second
brood of 3 d duration in order to increase the number of progeny from the treated males.

<u>Days 13 to 18</u>: Classify and count the eclosed flies with
respect to sex, eye shape, eye color and body color. Enter
the data in the following table.

<u>Result</u>

The rate of sex chromosome losses (L) is:

$$L\ (\%)\ =\ \frac{XO + PL}{XX + XY + XO + PL}\ \cdot\ 100$$

The progeny resulting from nondisjunction in the females are
not included in the calculation of the rate of losses.

<u>Literature</u>

GRAF, U., WüRGLER, F.E.: DNA repair dependent mutagenesis
    in <u>Drosophila melanogaster</u>. In: Genetics, Development,
    and Evolution of Drosophila (ed. S. Lakovaara). New
    York, London: Plenum Press 1982.
WüRGLER, F.E., GRAF, U.: Mutation induction in repair defi-
    cient strains of Drosophila. In: DNA Repair and Muta-
    genesis in Eukaryotes (eds. W.M. Generoso, M.D. Selby and
    F.J. de Serres). New York, London: Plenum Press 1980.
WüRGLER, F.E., FREI, H., GRAF, U.: Mutagen-sensitive mutants
    and chemical mutagenesis in Drosophila. In: Chemical
    Mutagens, vol. 10 (ed. F.J. de Serres). New York,
    London: Plenum Press 1986.

Fly count:

	XX ♀	XY ♂	XO ♂	PL ♂	PL ♂	ND ♀	Total (-ND)
Eye form Eye col. Body col.	+ + y	B $_w$(a) +	+ $_w$(a) y	B $_w$(a) y	+ $_w$(a) +	B/+/+ $_w$(a) +	
y w Control							
y w MMS							
mei-9^a Control							
mei-9^a MMS							

Rates of sex chromosome losses:

y w   Control :   ....

y w   MMS   :   ....

mei-9^a Control :   ....

mei-9^a MMS   :   ....

ZIMMERING, S.: Induced chromosome loss following treatment
    of postmeiotic cells of the Drosophila melanogaster male
    with MMS and DMN and mating with repair proficient fe-
    males and repair-deficient females mei-9^a and st
    mus-302. Mutation Res. 94, 79-86 (1982).
ZIMMERING, S., KAMMERMEYER, K.L.: On the nature of partial
    losses of the Y chromosome from treatment of ring-X/
    B^{S}Yy$^+$ males with diethylnitrosamine (DEN) or procar-
    bazine and matings with repair-deficient st mus302 fe-
    males of Drosophila. Mutation Res. 104, 121-123 (1982).

5.4 Somatic Mutation and Mitotic Recombination

Substantial parts of the body of the adult fly are produced
during larval and pupal development from the imaginal discs
(see Figures 9 and 22).

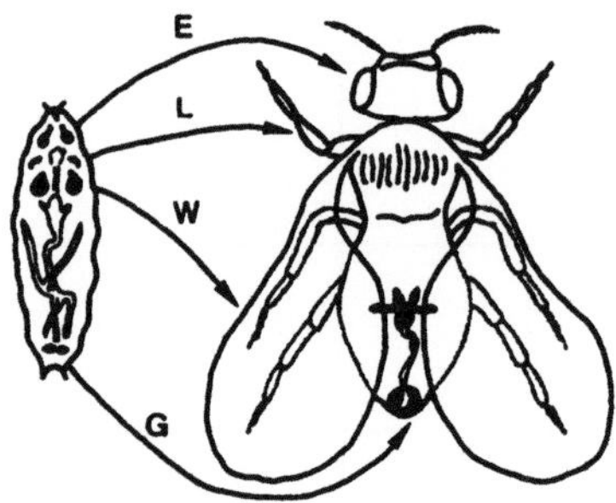

<u>Figure 22</u>. Some organs of the adult fly that develop from
imaginal discs (selection). Adapted from Markert and
Ursprung (1974). E = eye, G = genitalia, L = leg, W = wing.

One expects individual areas of the adult structures (e.g.
eye or wing) to be clones of progeny cells derived from one
single imaginal disc cell. This expectation can be tested
experimentally if it is possible to mark genetically individual clones during development in a stable way. One way in
which this can be done is by induced mutation. Drosophila
also offers the advantage of a second mechanism leading to
marked clones. In Drosophila somatic pairing of homologous
chromosomes takes place during mitosis; it is thus easy to
mark clones in heterozygous flies by induced mitotic recombination. This can be illustrated conveniently with wing
cell markers:

mwh/mwh : multiple wing hairs on each individual cell
            instead of a single hair
flr/flr : short and malformed wing hairs instead of long
            slender ones.

Figure 23 shows how a twin spot is produced by mitotic recombination and how a single spot is produced in the case of a
terminal deletion (partial chromosome loss).

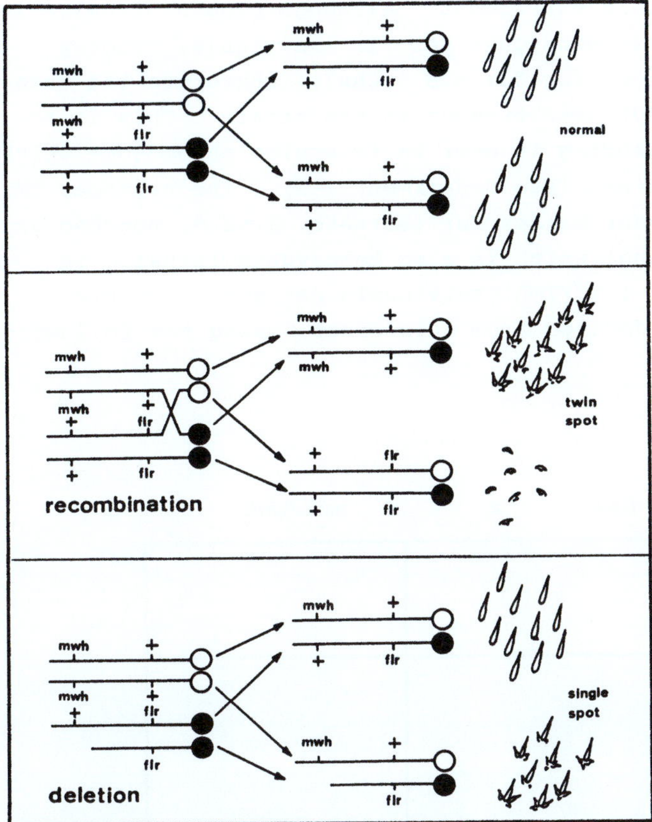

Figure 23. Results of normal chromosome segregation, mitotic recombination and terminal deletion in heterozygous (mwh + / + flr) wing imaginal disc cells.

Problem 1: Wing imaginal disc

Analyze the frequency and the size of single and twin spots on wings of flies from an untreated control and on wings of flies which were irradiated as larvae with 15 Gy X-rays at an age of 72 h.

Material

To produce the desired heterozygotes, we need virgin females from the flr^3/TM3, Ser strain (flr, flare, 3-38.8). The males are taken from an mwh strain (mwh, multiple wing hairs, 3-0.0). Of the flr alleles available, the flr^3 allele is best suited for this experiment. Although clones of cells in the wing which are homozygous for flr are viable, zygotes which are homozygous for flr are lethal; therefore the mutation has to be kept heterozygous in the strain. This is accomplished by balancing it over an inversion chromosome which eliminates crossovers (see experiment 7.2). The balancer TM3 carries the dominant marker Ser (Serrate, 3-92.5, notched at the tip of the wing) which is also homozygous lethal. In this way only the flr^3/TM3 individuals can survive. Predict the progeny derived from this cross, using the following table.

<u>Cross</u>

♀ flr^3/TM3, Ser   x   ♂   mwh/mwh

♂ ╲ ♀		

<u>Procedure</u>

<u>Day 1</u>: About 20 virgin mwh females and at least as many flr^3 males are crossed per culture bottle. Several bottles are set up for the collection of large numbers of larvae.
<u>Day 2</u>: After 24 h the parental flies are removed from the bottles (they can be used again for collection of more eggs).
<u>Day 4</u>: The larvae are collected from the culture bottles by pouring a 20% (w/v) sodium chloride solution (or 20% sucrose solution) into the bottles, swirling well and pouring the contents into a glass or plastic separation funnel (approx. 1000 ml, opening of the outlet 4-6 mm) which is fixed on a stand. The larvae now float on the surface of the saline solution. Open the stop cock and let the medium flow out (if

necessary break up small clumps with a glass rod). Most of
the larvae will adhere to the walls of the funnel. If the
larvae are not yet clean enough, add fresh saline solution
and repeat the procedure once more. Close the stop cock and
add water until the larvae start to sink. Open the stop cock
again and let the water flow out catching the larvae on a
fine nylon gauze. They can be washed again on the gauze if
needed. The larvae are irradiated with 15 Gy X-rays (45 keV,
10 mA, dose rate 20 Gy/min). Take care not to let the larvae
dry out. After the irradiation they are portioned out into
vials with standard medium. Larvae can be scooped up gently
on the end of a small spatula. A volume of about the size of
the head of a match is dropped into each vial. One group is
not irradiated; it serves as an untreated negative control.
<u>Days 11 to 13</u>: Collect the eclosed flies; only individuals
with normal (= non-Serrate) wings are kept. The wings can be
mounted immediately. The whole flies can also be stored for
extended periods in 70% ethanol in small scintillation vials.

Preparation of the wings

<u>Material</u>: Fine watchmaker's forceps, fine brush, needle,
microscopic slides, coverslips, distilled water, Faure's solution, small weights (metal cubes of approx. 1 cm length of
side).

<u>Faure's solution</u>:

distilled water	50 ml
gum arabic	30–40 g
glycerol	20 ml
chloral hydrate	50 g

The chloral hydrate is first dissolved in water without heating. Then the glycerol is added and mixed well. The gum
arabic is broken up into small pieces which are suspended in
the solution in a small gauze bag until all gum is dissolved. Avoid letting the gum come in contact with air during the dissolving process. Do not filter.

Freshly collected flies or flies which have been stored in
70% ethanol are put into distilled water in a small container. A drop of Faure's solution is put on a slide and one fly
transferred from the water into this drop. This is best done

with a fine brush. Avoid touching the wings with hard ob-
jects. Under a stereomicroscope the wings are now separated
from the fly's body with the help of two pairs of fine forceps (or a needle and forceps). The wings are then placed on
a clean slide by holding them only at the wing base. On the
slide the wings are spread out and arranged in rows without
touching the wing surface. Up to 20 pairs of wings can be
arranged on one slide. Let the slide dry in a dust-free atmosphere for 24 h. Put a small droplet of Faure's solution
in the center of a coverslip. Invert the coverslip and lower
it slowly onto the wings so that no air bubbles are formed
between the wings. Weigh the coverslip down with several
small weights (total weight ca. 400 g) and let dry for at
least 24 h. Coverslips of the dry preparations can be sealed
around the edges with clear nail polish.

<u>Analysis</u>

The wings are examined under a compound microscope at 400x
magnification. Both surfaces of the wing can be seen with a
slight focus change and are scored simultaneously for the
presence of mutant mwh or flr clones. Record single spots
and twin spots separately and determine the size of each spot
by counting the number of cells exhibiting the mutant phenotype. Make drawings of some typical spots on the following
sketches.

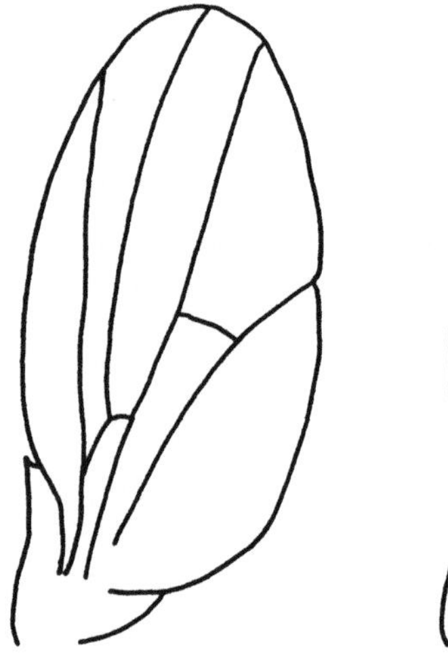 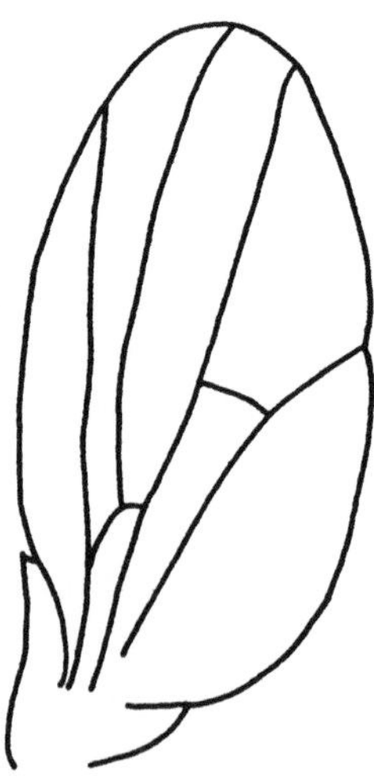 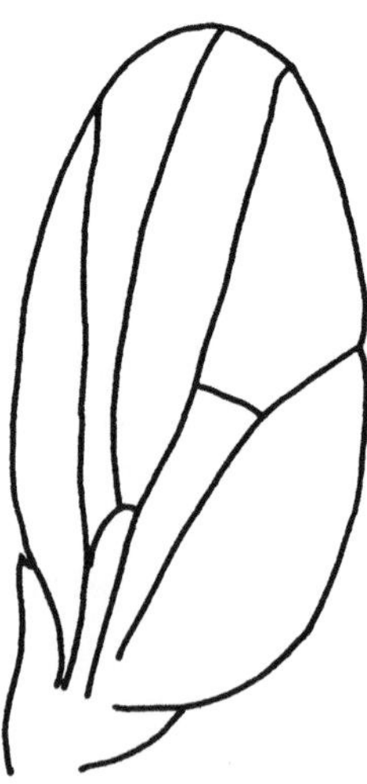

Experiment	Number of wings (w)	Single spots		Twin spots	
		s	s/w	t	t/w
Control					
X-rays					

Interpretation

What is the frequency (spots per wing: sp/w) of both types of spots in the two series? Are there differences with respect to the average size of the spots? Interpret these observations from a developmental point of view.

Problem 2: Eye imaginal disc

Analyze the frequency and the position of marked clones in the eyes of adult flies that are heterozygous for eye color markers and that were irradiated with X-rays in the first larval instar.

Material

Virgin females of the strain w^{CO} sn/w^{CO} sn; se/se and males of the strain w/w; se/se.
With respect to eye color the following three phenotypes can be distinguished:

w/w; se/se	white eyes
w^{CO}/w^{CO}; se/se	dark brown eyes
w^{CO}/w; se/se	intermediate between the above colors

For identification of clones that extend over the edge of the eye the sn marker on the X chromosome carrying w^{CO} is used:

sn^{+}/sn^{+}	normal bristles
sn/sn	singed bristles

Cross: ♀ w^{CO} sn/w^{CO} sn; se/se   x   ♂ w sn$^+$/Y; se/se

<table>
<tr><td>♂   ♀</td><td></td></tr>
<tr><td></td><td></td></tr>
<tr><td></td><td></td></tr>
</table>

Procedure

Day 1: Virgin w^{CO} sn/w^{CO} sn; se/se females are crossed with w sn$^+$/Y; se/se males in standard culture bottles.

Day 2: The parental flies are removed from the bottles after 24 h.

Day 3: After 12 to 24 h the larvae are collected from the bottles and irradiated with 12 Gy X-rays (for detailed instructions see problem 1).

Days 10 to 13: Collect the eclosed flies and analyze the eyes under a stereomicroscope at higher magnification (approx. 100x). Record white and dark brown spots, both singles and twins, and draw a few typical examples of spots in the lower half of the eye using the following sketches. Small white spots at the edge of the eye are not scored as they are very often due to nongenetic developmental disturbances.

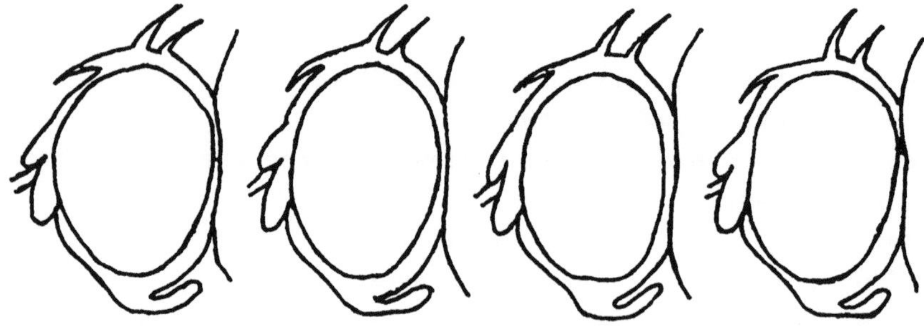

From a large number of eyes with spots analyzed by Becker
(1957) the following preferred position of the borders of the
spots in the lower eye half was determined; the eight sectors are numbered arbitrarily. Interpret this finding.

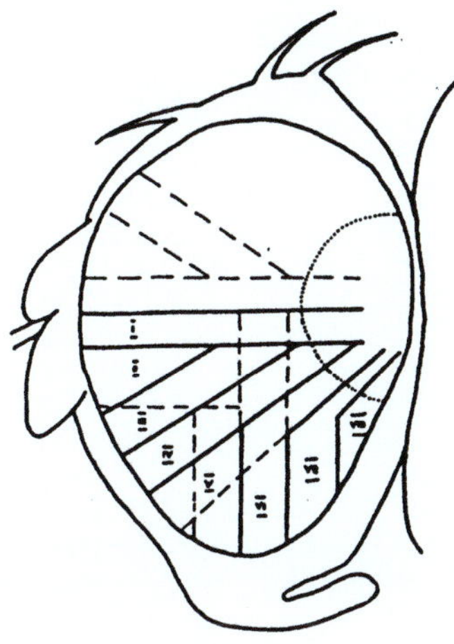

Literature

BECKER, H.J.: Mitotic recombination. In: The Genetics and
    Biology of Drosophila, Vol. 1c (ed. M. Ashburner and E.
    Novitski). New York, London: Academic Press 1976.
CAMPOS-ORTEGA, J.A., WAITZ, M.: Cell clones and pattern for-
    mation: Developmental restrictions in the compound eye
    of Drosophila. Wilhelm Roux's Archives <u>184</u>, 155-170
    (1978).
GEHRING, W.J. (ed.): Genetic Mosaics and Cell Differentia-
    tion. Berlin: Springer 1978.
GRAF, U., WüRGLER, F.E., KATZ, A.J., FREI, H., JUON, H.,
    HALL, C.B., KALE, P.G.: Somatic mutation and recom-
    bination test in <u>Drosophila</u> <u>melanogaster</u>. Environ.
    Mutagen. <u>6</u>, 153-188 (1984).
GRAF, U., FREI, H., KäGI, A., KATZ, A.J., WüRGLER, F.E.:
    Thirty compounds tested in the Drosophila wing spot
    test. Mutation Res. <u>222</u>, 359-373 (1989).
HAYNIE, J.L., BRYANT, P.J.: The effects of X-rays on the
    proliferation dynamics of cells in the imaginal wing disc
    of <u>Drosophila</u> <u>melanogaster</u>. Wilhelm Roux's Archives <u>183</u>,
    85-100 (1977).
MARKERT, C.L., URSPRUNG, H.: Entwicklungsbiologische Ge-
    netik. Grundlagen der modernen Genetik, Bd. 10.
    Stuttgart: Gustav Fischer 1974.
RANSOM, R. (ed.): A Handbook of Drosophila Development.
    Amsterdam: Elsevier/North Holland Biomedical Press 1982.
WüRGLER, F.E., VOGEL, E.W.: In vivo mutagenicity testing
    using somatic cells of <u>Drosophila</u> <u>melanogaster</u>. In:
    Chemical Mutagens, vol. 10 (ed. F. J. de Serres). New
    York, London: Plenum Press 1986.

6. Population Genetics

6.1 Determination of Allele Frequencies in Populations in Hardy-Weinberg Equilibrium

The Hardy-Weinberg law is a formulation of the relationship between the frequencies of genes in a population and frequencies of genotypes. The basic model describes the simplest nontrivial case, i.e. one autosomal locus with two alleles in a randomly mating (panmictic) diploid population in which nothing occurs to change the frequency of the two alleles. This means that the population is infinitely large (or at least large enough to eliminate the effects of chance) and that factors such as mutation, selection, and migration have no effect on gene frequency. The Hardy-Weinberg law states that if allele A has a frequency of p, and allele a a frequency of q, then after one generation of random mating the population will consist of the genotypes AA, Aa and aa in the frequencies p^2, 2pq and q^2, respectively, and furthermore that these frequencies will be maintained in further generations as long as the basic conditions hold. Such a population is said to be in Hardy-Weinberg equilibrium. We can see that p^2 or q^2 represents the probability that two gametes both carrying the same allele (A or a) will unite at fertilization. 2pq is the probability that an A gamete will unite with an a gamete. The basic model can be modified to describe sex-linked genes, multiple alleles, effects of selection, mutation, and genetic drift.

Problem

In Drosophila a gene is known which affects the body color. The dominant wild type allele (T) produces the normal light brown to gray coloring, the recessive allele t (= tan) results in a darker brown coloring. Note: Here for simplicity, we do not use usual Drosophila symbolism. T/T, T/t, and t/t animals have identical viability and frequency of reproduction. In a theoretical experiment it is possible to visualize the allele frequencies in a panmictic population in detail. We assume that the allele frequencies are identical among oocytes and sperm. For both types of gametes let the frequency of the T alleles be = p and that of the t alleles = q, and since there are only two alleles, p + q = 1.

Procedure

Using the following combination square work out genotypes and frequencies of each combination.

<table>
<tr><td></td><td colspan="2" align="center">Oocytes</td></tr>
<tr><td rowspan="2">♂   ♀</td><td>Genotype
Frequency</td><td>G
F</td></tr>
<tr><td>Sperm   G F</td><td></td><td></td></tr>
<tr><td>G F</td><td></td><td></td></tr>
</table>

The frequency of the different genotypes in the entire progeny is then:

T/T	&	T/t	&	t/t
....		....		....

The whole population of progeny is now bred with the frequencies of genotypes as determined above. We assume panmixia, which means that mating and reproduction within a population occur entirely at random. As a consequence, the probability for a specific mating between two genotypes is determined only by the respective frequencies in the population. We first work out what types of oocytes originate from the females of this population. The females of the T/T genotype, which occur with a frequency p^2 in the population, produce T oocytes only. The T/t females form half T oocytes and half t oocytes, whereas the t/t females give rise to t oocytes alone.

	Frequency	
	T oocytes	t oocytes
T/T females	....	....
T/t females	....	....
t/t females	....	....

These calculations can be done for all the sperm as well.

Show the **expected** progeny in the following checkerboard:

♂ \ ♀		

        T/T   &   T/t   &   t/t

....

These calculations can be repeated as many times as one
wants. The identical result demonstrates that according to
the Hardy-Weinberg law the ratio of the gametes (or genotypes, respectively) remains constant in a panmictic population.

Consider a population started from individuals of which 20%
were TT ♀, 20% TT ♂, 30% tt ♀, and 30% tt ♂. Is this
founder population in Hardy-Weinberg equilibrium? Work out
what the frequencies of the different genotypes would be
expected to be after one generation of random mating.

The Hardy-Weinberg equilibrium can also be used the other way
round to calculate gene frequencies where they are unknown.
By assuming that a population is in Hardy-Weinberg equilibrium, we can obtain the frequency of at least one genotypic
class, the homozygous recessive, even when dominance is present. The frequency of the recessive allele (q) is then simply the square root of the genotypic frequency of the homozygous recessive class (q^2). The frequency of the dominant
allele (p) is calculated as 1-q.

For situations where there is no complete dominance, the frequencies of the two alleles can be calculated directly from the three phenotypes in population samples:

$$p = \frac{D + H/2}{D + H + R}$$

where D = No. of homozygotes for one allele

H = No. of heterozygotes

$$q = \frac{R + H/2}{D + H + R}$$

R = No. of homozygotes for the other allele

These determined values can be used to calculate predicted values of p^2, 2pq, and q^2 which can then be compared statistically to the observed values of D, H, and R to establish whether or not the population in question is in Hardy-Weinberg equilibrium.

Literature

HARDY, G.H.: Mendelian proportions in a mixed population. Science 28, 49-50 (1908).
WEINBERG, W.: Ueber den Nachweis der Vererbung beim Menschen. Jahresverh. Verein vaterl. Naturk. Württemberg 64, 368-382 (1908).

6.2 Variations in Populations with Natural and Artificial Selection

In this experiment a fly population is bred over several generations. In each generation the composition of the population is determined. The experiment starts with the following cross (vg, vestigial wings, 2-67.0):

P : Genotype   ♀ vg^+/vg^+ x   ♂ vg/vg

     Phenotype   normal   vestigial wings

Indicate the expected F_1 genotype and phenotype below:

F_1: Genotype ....   Phenotype ....

The F_1 flies are intercrossed. Work out the progeny expected in the F_2 and record them below:

F_2:

Genotypes	Phenotypes	Expected frequencies

Without killing the F_2, count the numbers of normal and mutant individuals, ignoring sex, and enter the numbers in the table below. These flies are then used to establish the subsequent generation. This can be done in two different ways: Breeding without artificial selection (normal and mutant phenotypes are cultured) or breeding with artificial selection (only normal phenotypes are cultured).

Procedures

Experiment A: NATURAL selection:

Approximately 30 females and 30 males are put into a standard culture bottle in a proportion according to the progeny count. Example: 250 normal and 50 mutant flies were counted. In this case 25 normal and 5 mutant flies of each sex are taken.

Experiment B: ARTIFICIAL selection:

Approximately 30 _normal_ females and 30 _normal_ males are cultured in a bottle. For selecting against the mutant phenotypes all the homozygous mutant flies are discarded in each generation.

Results

Selection	Generation and date	Number of flies with phenotype		Total flies	Frequency of vg/vg
		normal	vestigial		
A NATURAL	F_2 F_3 F_4 F_5				
B ARTIFICIAL	F_2 F_3 F_4 F_5				

Draw a graph of your results:

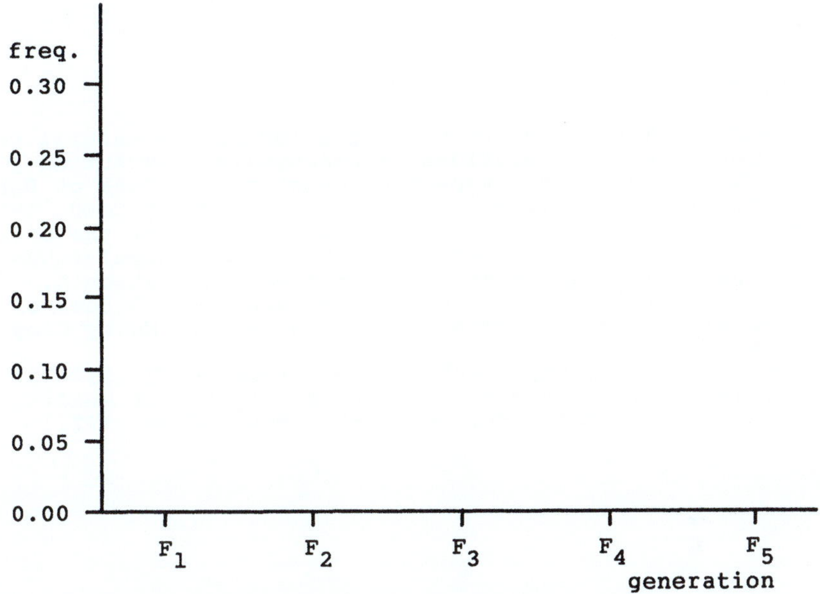

Questions

Does it seem likely that the three genotypes in these populations have equal viability?
Do you think that a recessive mutant could ever be completely eliminated from a population by selection against the homozygote?

Population modeling

For evaluating such an experiment one proceeds as follows: A simple mathematical model is set up given specific assumptions. From this model the changes to be expected in the population are calculated, and this prediction is compared with the experimental result. If the prediction and the result do not coincide, then specific assumptions in the model must be changed. A new expectation can be calculated and compared again with the experiment. This procedure is re-

peated until a model is found which describes the results
with a desired precision.

Literature

DOBZHANSKY, T.H.: Adaptive changes induced by natural selec
    tion in wild populations of Drosophila. Evolution $\underline{1}$,
    1-16 (1947). In: Papers on Genetics. A Book of Read-
    ings (ed. L. Levine). St. Louis MO: Mosby Comp. 1971.
KERR, W.R., WRIGHT, S.: Experimental studies of the distri
    bution of gene frequencies in very small populations of
    <u>Drosophila</u> <u>melanogaster</u>. I. Forked. Evolution $\underline{8}$,
    172-177 (1954). In: Papers on Genetics. A Book of
    Readings (ed. L. Levine). St. Louis MO: Mosby Comp.
    1971.
STERN, C.: The Hardy-Weinberg law. Science $\underline{97}$, 137-138
    (1943). In: Papers on Genetics. A Book of Readings
    (ed. L. Levine). St. Louis MO: Mosby Comp. 1971.

6.3 Mating Behavior

The successful mating of flies, which is presupposed in all
our crosses, actually requires a complicated, genetically
determined courtship ritual. We shall observe the normal
mating behavior of flies before analyzing changes in this
behavior.

Problem

To determine the sequence and the individual elements of the
different activities in the mating behavior.

Procedure

Put one pair of virgin wild type flies into a container of
about 2 cm diameter with a transparent cover. Observe the
behavior of the female and the male under a stereomicroscope
at low magnification (10x). Record the observations in the
table below.

Result

Time	Female	Male

Literature

EHRMAN, L.: Sexual Behavior. In: The Genetics and Biology
    of Drosophila, vol. 2b (eds. M. Ashburner and T.R.F.
    Wright). New York: Academic Press 1978.
GAILEY, D.A., JACKSON, F.R., SIEGEL, R.W.: Male courtship in
    Drosophila: The conditioned response to immature males
    and its genetic control. Genetics 102, 771-782 (1982).
HOTTA, Y., BENZER, S.: Courtship in Drosophila mosaics: sex
    specific foci for sequential action patterns. Proc. Nat.
    Acad. Sci. (USA) 73, 4154-4158 (1976).

6.4 Mating Preferences

As we have seen in a previous experiment, the Hardy-Weinberg
law is only applicable under the assumption of panmixia in a
population. However, it can be shown that for specific alleles in a population the frequency of the carriers of this
allele is different from the expectation based on the frequency of the allele in the population. In other words, mating preferences exist within a population. This phenomenon
can be demonstrated with the help of two different monohybrid
crosses. The mutant yellow is best suited for this. This
mutation causes not only a yellow body color, but also leads
to a changed behavior of the males in the courtship ritual
that precedes the mating.

<u>Problem</u>

Is there a difference between the mating success of wild type
males and of yellow males?

<u>Material</u>

Virgin 4-day-old females of a wild type strain and 2-day-old
males of the wild type as well as of the yellow strain (y,
1-0.0). The males are kept in sexual isolation for at least
1 day.

<u>Procedure and evaluation</u>

<u>Day 1</u>:   The following two crosses are set up:
           ♀ wild type   x   ♂ wild type
           ♀ wild type   x   ♂ yellow
For each cross at least 50 vials with one virgin female and
one male each are made. The vials with the pairs of flies
are left undisturbed in the dark for <u>3 hours</u>. At the end of
the mating period each vial is emptied separately, the male
is discarded and the female returned to the vial where she is
left until the end of the experiment.
<u>Day 10</u>: A criterion used for a successful mating within the
3-hour period, is the deposition of inseminated eggs capable

of development by the individual females. For the scoring it
is sufficient to record only whether or not there are progeny
in a vial; the flies need not be counted.
What is the percentage of vials with progeny in the two
crosses?

Cross	Total no. of vials	No. vials with progeny	% vials with progeny
y^+/y^+ x y^+/Y			
y^+/y^+ x y/Y			

Conclusions

What can you conclude from your results about the effectiveness of mating of the two types of males? List as many hypotheses as you can think of to explain your results.

Literature

AVERHOFF, W.W., RICHARDSON, R.H.: Pheromonal control of
    mating pattern in Drosophila melanogaster. Behavior Ge-
    netics 4, 207-225 (1974).
BASTOCK, M.: A gene mutation which changes a behavior pat-
    tern. Evolution 10, 421-439 (1956).
DOW, M.A.: The genetic basis of receptivity of yellow mutant
    Drosophila melanogaster females. Behavior Genetics 6,
    141-143 (1976).
HARDELAND, R.: Lighting conditions and mating behavior in
    Drosophila. American Naturalist 105, 198-200 (1971).
MAGALHAES, de L.E., RODRIGUES-PEREIRA, M.A.Q.: Frequency de-
    pendent mating success among mutant ebony of Drosophila
    melanogaster. Experientia 32, 309-310 (1976).
STROEMNAES, O., KVELLAND, I.: Sexual activity of Drosophila
    melanogaster males. Hereditas 48, 442-470 (1962).

7. Cytology and Cytogenetics

7.1 Microscopic Analysis of Mitotic and Polytene Chromosomes

Two tissues from the last larval instar (L3) are suited for cytological analysis: (1) The neural or cerebral ganglion (= supraesophageal ganglion) for mitotic chromosomes and (2) the salivary glands for giant chromosomes.

<u>Preparation of tissues</u>

To obtain satisfactory results, the optimal rearing of the larvae is important. The parental flies are left in the culture bottles for egg deposition for not more than 24 h. After discarding the parents, fresh live yeast is added to the bottles as additional food. Third instar larvae are taken from uncrowded cultures. For the preparation of especially large salivary gland chromosomes use larvae grown at low temperature (below 18°C). The larvae are raised in bottles without paper and are removed with a brush when they start to leave the medium and crawl up the walls of the bottle towards the end of the third larval instar. Figure 24 shows how a larva is held with two needles or two fine forceps (e.g. watchmakers' forceps Dumont No. 5) in a drop of Drosophila Ringer solution and how the anterior part with the mouth hooks is pulled away with a sudden movement. The posterior two thirds of the larva can be severed with a razor blade and discarded. Both the neural ganglion and the two salivary glands with the joint opaque fat bodies are found in the tissues that are attached to the anterior third of the larva. With a fine needle and forceps the tissues are teased out under intermediate magnification (20-25x). The neural ganglion and the attached imaginal discs are separated; the imaginal discs need not be removed. The salivary glands are gently pulled free of the fat bodies which are discarded. The larger cells lie in the center and at the posterior end of the salivary gland which contains approximately 100 to 200 cells.

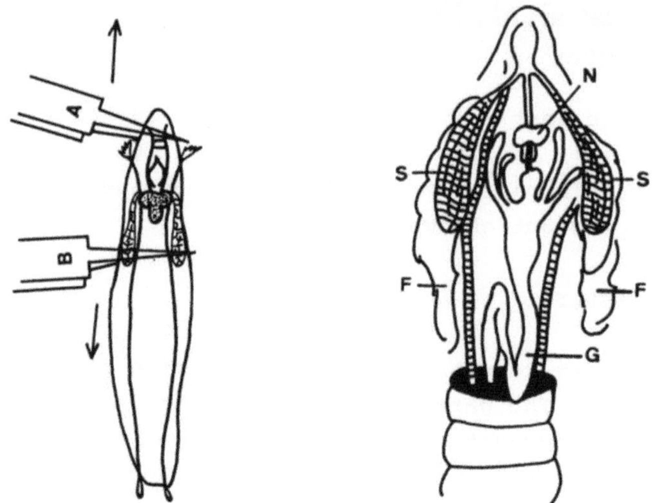

Figure 24. Preparation of larvae. After Ransom (1982) and Schlösser (1971). F = fat bodies, G = gut, N = neural ganglion, S = salivary glands.

Preparation of siliconized glassware

Siliconized glassware is needed to prevent tissue from sticking to glass surfaces, e.g. in the preparation of mitotic chromosomes. The easiest way to siliconize slides or coverslips is to immerse them for a few minutes in silicon oil and to wipe them clean with tissues. (Pipettes used for oil should be rinsed with alcohol.)

Recipes

(a) Isotonic physiological salt solutions:
Drosophila Ringer: 6.7 g NaCl in 1000 ml distilled water.
Shen solution: 9 g NaCl + 0.42 g KCl + 0.25 g $CaCl_2$ in 1000 ml distilled water.

(b) Hypotonic sodium citrate solution:
1 g $Na_3C_6H_5O_7 \cdot H_2O$ in 100 ml distilled water.

(c) Staining solutions:

<u>Aceto-Orcein</u>: For immediate use, not permanent: Pour 50 ml
boiling glacial acetic acid over 1 g Orcein, let cool down,
add 50 ml distilled water, filter. Permanent solution: Dissolve 2.2 g Orcein in 100 ml boiling glacial acetic acid;
dilute 1 : 1 with distilled water and filter only immediately
before use.

<u>Orcein-Fast Green</u>: Equal volumes of glacial acetic acid and
85% lactic acid are mixed. 2 g natural Orcein and 0.25 g
Fast Green are dissolved in 100 ml of the solvent mixture
without heating. The staining solution is filtered several
times before use.

<u>Giemsa</u>: One uses commercial solutions and dilutes them before use with phosphate buffer according to manufacturer's
instructions. Do not filter! Keep solution in the dark!

(d) Fixative:

3 parts absolute ethanol plus 1 part glacial acetic acid.
Make up fresh for use.

(e) Phosphate buffer according to Sörensen:

Mix 50 ml of a 1/15 M Na_2HPO_4 with 50 ml of a 1/15 M
KH_2PO_4 solution. This gives a pH 6.8 buffer.

(f) Colchicine:

Use at least an 0.2% solution (0.2 g colchicine in 100 ml
Shen solution).

Problem 1: Mitotic chromosomes of the neural ganglion

Problem

Prepare squashes of mitotic chromosomes of the neural ganglion of a wild type strain, stain them and identify the individual chromosomes by microscopic analysis of the preparation.

Procedure

1. For the accumulation of metaphases, transfer the neural

ganglion (with or without attached imaginal discs) into a drop of Shen solution with 0.2% colchicine on a well slide for 60 min.

2. Using a siliconized pipette containing a small amount of sodium citrate, transfer the ganglion into 2 to 3 drops of sodium citrate solution (hypotonic shock) on a siliconized slide for 10 min.

3. Transfer the ganglion into a drop of 50% acetic acid on a siliconized slide for 15 min.

4. Using a siliconized pipette transfer the ganglion into a fresh drop of 50% acetic acid on a nonsiliconized slide. A siliconized coverslip is then added and covered with a piece of filter paper. The preparation is then squashed with strong thumb pressure. Apply only vertical pressure; sideward movements destroy the preparation! The degree of squashing can be controlled under a phase-contrast microscope.

5. When the squash appears satisfactory, freeze the preparation on a piece of dry ice kept in an insulated container (e.g. wooden box lined with corrugated paper or styrofoam container). This takes at least 3 min.

6. While the slide is still lying on the dry ice, the coverslip is flipped off with a pre-cooled razor blade.

7. Thaw the preparation by immersing it in an ethanol-acetic acid fixative for 5 min or longer. The quality of the preparation is improved by leaving it in the fixative overnight. Ethanol-acetic acid fixative gives a more transparent cytoplasm than absolute alcohol.

8. Remove the slide from the fixative and add one drop of staining solution (Orcein-Fast Green or Giemsa) and cover with a clean, nonsiliconized coverslip.

9. Absorb the excess fluid with filter paper and seal with clear nail polish along the edges of the coverslip to avoid desiccation.

10. The preparation is then viewed under a compound microscope, if possible with phase-contrast optics. It can also be dehydrated and mounted as a permanent preparation (before putting on the coverslip!).

11. Identify the acrocentric X chromosome, the small Y chromosome with the two unequal arms, the metacentric chromosomes 2 and 3 as well as the dot-like chromosomes 4. Make a drawing of a metaphase. Did you analyze a female or a male larva?

Problem 2: Polytene chromosomes of the salivary gland

Problem

Prepare squashes of polytene chromosomes of the salivary
gland of a wild type strain, stain them, and analyze the
chromosomes and their banding pattern.

Procedure

1. The dissected salivary glands from one larva are put into
a drop of Aceto-Orcein staining solution. The preparation is
covered with a coverslip and left for about 3 min. The duration of the staining can be modified with subsequent preparations depending on the effectiveness of the staining solution.
2. The staining is terminated by soaking up the stain under
the coverslip with a strip of filter paper.
3. The coverslip is now covered with a broad piece of filter
paper and the preparation is squashed so that the cells and
nuclear membranes burst and the chromosome arms are spread.
The simplest way is application of thumb pressure or tapping
with a suitable instrument (e.g. wooden handle of a dissecting needle).
4. To avoid desiccation of the preparation it is sealed
along the edges of the coverslip with clear nail polish.
5. Analyze the polytene chromosomes under a compound microscope. Observe and draw:
(a) The central heterochromatic area (chromocenter), which
contains the centromeres (kinetochores).
(b) Try to find the short chromosome 4.
(c) Try to identify specific chromosome ends. Sketch the
banding pattern observed under the microscope and compare it
with the schematic drawings in Figure 25.

Literature

HARSHMAN, L.G.: A technique for the preparation of Droso-
    phila salivary gland chromosomes. Drosophila Inf. Serv.
    52, 164 (1977).
KING, R.C. (ed.): Handbook of Genetics, vol. 3: Inverte-
    brates of Genetic Interest. New York: Plenum Press 1975.
OSTER, I.I., BALABAN, G.: A modified method for preparing
    somatic chromosomes. Drosophila Inf. Serv. 37, 142-144
    (1963).

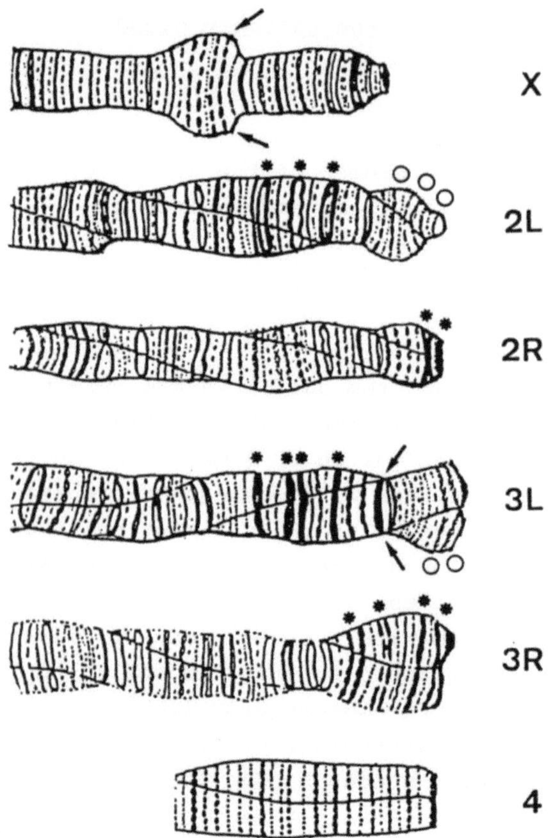

<u>Figure 25</u>. Ends of the chromosome arms with banding pattern
in salivary gland chromosomes. After King (1975). Arrow:
characteristic shape, asterisk: prominent bands, ring: no
bands or only weakly staining.

PAINTER, T.S.: A new method for the study of chromosome re-
    arrangements and plotting of chromosome maps. Science
    <u>78</u>, 585-586 (1933). In: Classic Papers in Genetics (ed.
    J.A. Peters). Englewood Cliffs NJ: Prentice-Hall 1959.
RANSOM, R.: Techniques. In: A Handbook of Drosophila Deve-
    lopment (ed. R. Ransom). Amsterdam: Elsevier Biomedical
    Press 1982.
SCHLÖSSER, K.: Experimentelle Genetik. Heidelberg: Quelle
    und Meyer 1971.
SWANSON, C.P.: Cytology and Cytogenetics. London: MacMil-
    lan 1963.
SWANSON, C.P., MERZ, T., YOUNG, W.J.: Cytogenetics. The
    Chromosome in Division, Inheritance and Evolution, 2nd
    ed. Englewood Cliffs NJ: Prentice-Hall 1981.

7.2 Balancer Chromosomes

In Drosophila genetics balancer chromosomes are used for the
elimination of various undesired consequences of meiotic
crossing over events. With the help of balancer chromosomes
multiply marked mutants, which would be inviable or sterile
in homozygous condition, can be cultured for many generations
without problems (balanced strains). This is especially true
for recessive lethals. In addition, balancer chromosomes are
indispensable in the construction of new strains. They are
used to propagate multiply marked chromosomes via heterozygous females without the desirable combinations being broken
up from one generation to the next by crossing over. Heterozygous males in which no meiotic crossing over takes place
are no problem in this respect when constructing new
strains. The use of balancer chromosomes goes back to an
early observation by drosophilists that certain types of
chromosomes in heterozygous combination with marked chromosomes were able to reduce or even practically eliminate the
frequency of crossing over. It was postulated that these
chromosomes contain a crossover suppressor (symbolized by
"C"). In later investigations it was shown that C-carrying
chromosomes contain large, often complex inversions (Sturtevant 1926). In heterozygous flies the inversions can be analyzed cytologically in salivary gland chromosomes. Two problems will help to work out the mechanism of action of balancer chromosomes.

Problem 1: Analysis of inversion loops

In the salivary gland cells of inversion heterozygotes the
somatic pairing of the homologous chromosomes leads to the
formation of a loop in the inverted region. This is the only
way in which each region of a chromosome is able to pair with
its homolog. In the light microscope the continuity of a
chromosome can be followed with the help of the banding pattern. For simplification the bands can be symbolized by let-
ters.

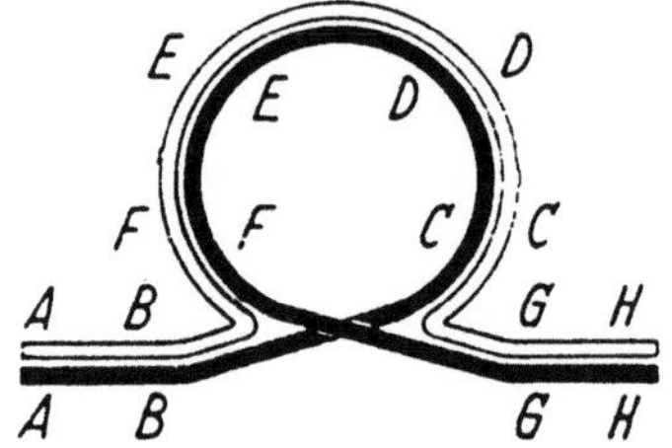

The normal chromosome has the sequence:
    A B C D E F G H.
In contrast to this the inversion chromosome is:
    A B F E D C G H.
The inverted region extends from C to F. Formally the inverted chromosome can be reverted back into a normal chromosome if one takes the two break points B.F and C.G and "turns
around" the inversion region.

Using this approach analyze the following loop drawings. All
three are the consequence of two separate inversions. What
is the relative position of the two inversions in each case?

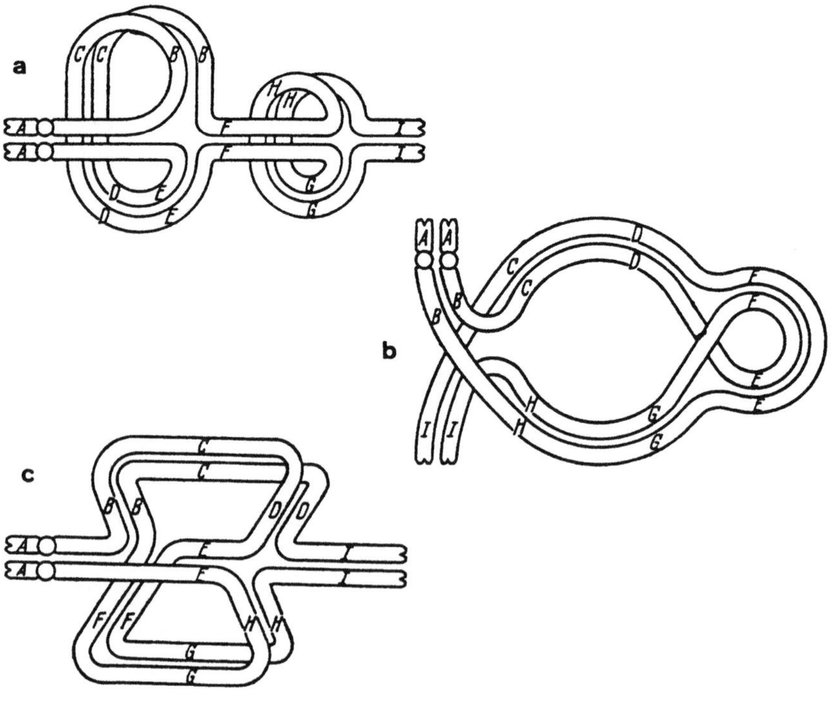

Problem 2: Crossing over products in inversion heterozygotes

In meiosis of female inversion heterozygotes the chiasmata
appearing in the tetrad stage indicate the crossovers which
took place during an earlier stage of meiosis.
Analyze the consequences of crossing over events occurring in
the inversion loop. Draw the four chromatids with the loci
labelled as they will appear at early anaphase I (after separation of the centromeres). Be sure to include all 8 chromatid ends.

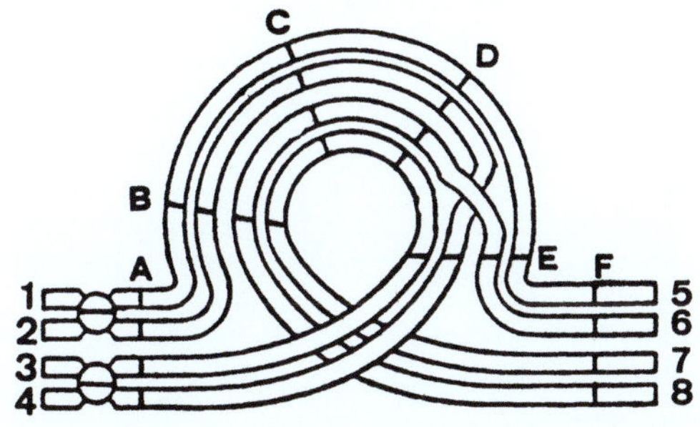

What will happen subsequently and during the second meiotic
division in such a case? What are the expected consequences
with respect to fertility of such female inversion heterozygotes? Would the inversion have any effect on the recovery
of crossover products in females homozygous for it? What are
the genetic consequences? Suggest a better term than "crossover suppressor" to describe the action of the inversion.

Experimental analysis

Using the techniques described in experiment 7.1 examine and
sketch the salivary gland chromosomes of the strains Cy/Pm ;
H/Sb or those of female progeny of the cross Basc ♀ x wild
type ♂. Would you expect to find inversion loops (a) in
homozygous Basc females or (b) in Basc males?

Literature

BEADLE, G.W., STURTEVANT, A.H.: X-chromosome inversions and
    meiosis in _Drosophila melanogaster_. Proc. Nat. Acad.
    Sci. (USA) 21, 384-390 (1935).
CARSON, H.L.: The selective elimination of inversion dicen-
    tric chromatids during meiosis in the egg of _Sciara im-
    patiens_. Genetics 31, 95-113 (1946).
RANSOM, R. (ed.): A Handbook of Drosophila Development.
    Amsterdam: Elsevier Biomedical Press 1982.
RIEGER, R., MICHAELIS, A., GREEN, M.M.: Glossary of Genetics
    and Cytogenetics. Berlin: Springer 1976.
STRICKBERGER, M.W.: Genetics, 2nd ed. New York: MacMillan
    1976.
STURTEVANT, A.H.: A crossover reducer in _Drosophila melano-
    gaster_ due to inversion of a section of the third chromo-
    some. Biol. Zentralbl. 46, 697-702 (1926).

8. Molecular Biology

8. Molecular Biology of Drosophila: An Overview

The foregoing chapters of this book describe classical genetic experiments and analyses making use of Drosophila as an
easy and inexpensive experimental animal. Its short generation time, ease of culture, large numbers of progeny, small
number of chromosomes and easily recognizable genetic traits
have led to its being the most thoroughly genetically described multicellular eukaryote with thousands of mutants and
chromosomal aberrations described and mapped.

An organism that is well known genetically is also attractive
to the molecular biologist for the application of the sophisticated techniques being used to unravel the ultimate secrets
of the structure and function of the genetic material from
the molecular to the population level.

Modern molecular biology requires laboratories equipped with
complex and expensive instruments. It also makes use of expensive reagents and requires considerable technical exper-
tise from its practitioners. We do not intend, therefore, to
give detailed instructions for particular experiments to be
done with large numbers of students as we did in previous
chapters, but will rather concentrate on short summaries of
some of the major approaches used and a few examples from
recent Drosophila work. The references included will allow
those interested to find descriptions of relevant methods.

Studies of DNA

The blueprint for a Drosophila fly, like that of any other
organism, is encoded in the sequence of nucleotides in DNA.
The DNA, in turn, specifies RNA which specifies protein
structure (students are referred to genetics and molecular
biology textbooks for details). Much of molecular biology
therefore involves the extraction, characterization and manipulation of DNA. Drosophila DNA may be extracted from
adult flies, larvae, embryos, isolated organs or cells grown
in culture. Cell and tissue culture in Drosophila have

proved rather difficult. Embryo-derived cells can be cultured and have found some practical use for the growth of certain insect viruses (King, 1975; Roberts, 1986; Ashburner, 1989a).

Detailed techniques for DNA extraction from Drosophila are given in Roberts (1986) and Ashburner (1989a). Techniques are now so refined that the DNA from a single fly can be extracted for study. Total DNA may be extracted or nuclear and cytoplasmic (mitochondrial) fractions separated. Mitochondrial DNA, which is inherited maternally, is highly conserved in evolution and provides an interesting subject for studies of related races and species. Complete DNA replacement of mitochondrial DNA has recently been accomplished in Drosophila (Niki et al., 1989). This should make it a promising organism for studies of nuclear-cytoplasmic interactions.

Once DNA has been extracted, it can be characterized and manipulated. One type of chemical investigation, for example, is the determination of the GC content. DNA dissociation-reassociation studies can be performed in order to determine the complexity of the genome.

Techniques for the further study of DNA usually require that the long molecules present in chromosomes be cut into manageable shorter pieces. This is accomplished by so-called restriction enzymes. These enzymes, isolated from microorganisms, recognize particular nucleotide sequences, and each one cuts DNA at its own recognition site wherever that sequence occurs. This generates a characteristic series of fragments of different lengths which can then be separated by gel electrophoresis.

Studies of the same segments of DNA (defined by a specific DNA probe) from different individuals show that DNA contains many slight variations in nucleotide sequence. If a change, deletion, or rearrangement affects the sequence recognized by a particular restriction enzyme, then the enzyme will no longer cut at that location and a fragment of a different length will result. Such restriction fragment length polymorphisms (RFLPs) can be used to estimate the variability within populations, for gene mapping (see Alberts et al., 1989, p. 182) and in investigations of chromosomal regions of interest. A recent example of work of this type is a new study of the Bar gene in Drosophila. Classical cytogenetic analyses had indicated that the Bar mutation results from a duplicated region of the X chromosome. By using restriction enzymes to study and map the region around the break-point of this rearrangement in combination with sequence data, Tsubota et al. (1989) have come to the conclusion that the insertion of a transposable element at this location was responsible for the rearrangement.

Since it is not easy to study single molecules of DNA, many types of studies require some method of making copies of a single molecule (amplification). The polymerase chain reaction (PCR) is an in vitro technique for amplification of DNA molecules of up to 2000 base pairs (Saiki et al., 1988). Another method is to insert the DNA of interest into a DNA vector molecule, usually a bacterial plasmid or a virus genome or a transposable element which is in turn introduced into cells (usually bacteria) under conditions which lead to its replication. This is known as gene cloning (see Ashburner, 1989a, for details of cloning vectors).

Most restriction enzymes cut the DNA in such a way that single stranded sticky ends are left. This means that the single stranded overhang on the one end of the molecule is complementary to the single stranded overhang on the other end. Fragments from any DNA cut in this way with the same enzyme can then be joined together by allowing their sticky ends to anneal by base pairing and then using a ligase enzyme, to join the sugar-phosphate backbones covalently. The vector molecules are then introduced into cells, and marker genes are used in ingenious ways to identify by selection or screening procedures the cells which contain vectors with inserts (see Brown, 1986, for a general discussion). If total DNA is extracted, cut up and fragments inserted into vectors and cloned, and if enough pieces have been used, it is expected that every gene should be present in at least one cell. These cells are allowed to grow into separate colonies, and such a collection of cloned fragments is known as a gene library. Gene libraries have been constructed for most genetically studied organisms including Drosophila.

Special techniques also exist for cloning specific parts of the genome (e.g. the Y chromosome) or specific genes. One way of separating one single gene from the many thousands in the genome is by means of transposon tagging. A transposable element of known sequence (such as the Drosophila P element, see experiment 4.8) is used under conditions allowing it to transpose in a genome in which a loss of function of the searched-for gene can be recognized phenotypically. If such a mutation occurs due to the insertion of the transposon into the gene, DNA from the mutant can be hybridized with DNA from the transposon. This allows the location of any site containing the transposon. Such sites are then extracted with the DNA surrounding them and cloned and should include the mutated gene of interest (see for example Rubin and Spradling, 1983).

Once a gene has been cloned, its DNA can be sequenced (see Maxam and Gilbert, 1977, and Sanger et al., 1977 for techniques). Information derived from sequencing is stored in computer data banks such as that of the BIONET electronic network, funded by the U.S. National Institutes of Health. Computer programs are available that allow the identification of varying degrees of homology in DNA sequences or deduced protein sequences (Kristofferson, 1987). Sequencing studies have led to the determination of the general structures of genes of different types, to the identification of particular short sequences that act as signals for replication, the start and end of transcription, the chemical modification of DNA and many sequences that provide recognition sites for various enzymes and regulatory proteins. Since the primary sequence of nucleotides in RNA is determined by the sequence in DNA, comparisons of DNA and RNA sequences shed light on the various types of post-transcriptional modifications that occur in higher organisms. The particular sequences that are characteristic of the ends of transposons and viral genomes capable of inserting into the genome of other organisms have been identified as well as those of chromosome structures such as centromeres and telomeres. Such studies have also helped to identify families of identical or similar genes.

Comparisons of sequences of similar genes from related species can help in the understanding of how genes may evolve at
the molecular level, and the comparison of alleles elucidates
the nature of mutation (see for example Pastink et al.,
1989). A surprising homology recently discovered is that
between the Drosophila developmental gene wingless and the
mouse mammary oncogene int-1 (Rijsewijk et al., 1987). Many
Drosophila genes have been sequenced. Selected examples are
listed in the following table.

One example of a particular sequence of great interest, first
recognized in Drosophila, is the so-called homeobox, a 180
base pair sequence that was found in different genes regulat-
ing the control of pattern formation in Drosophila embryos.
Subsequently, homologous sequences have been found in a broad
range of vertebrates as well as in invertebrates (see review
by Gehring and Hiromi, 1986).

Machines are now available which can synthesize chemically
short pieces of DNA of desired sequence. These can be introduced into native DNA molecules to produce site-specific mutations, lead to new restriction sites and other variations
limited only by the imagination and ingenuity of the experi-
mentalists.

Understanding how genes are structured and how the various
kinds of molecular signals work has allowed genetic engineering in the true sense of design engineering. Regulatory
signals from one organism can be spliced to protein-specify-
ing sequences from another and introduced into the genetic
material of still a different organism.

Once a particular gene has been cloned it can be introduced
into cells, and if it contains all the signals needed for
correct functioning, will produce its product. It has been
one of the hopes of modern genetics that this approach could
be used for the correction of genetic disease. This has been
accomplished in Drosophila. One example was the correction
of the abnormal behavior pattern of flies carrying the mutant

Selected examples of cloned Drosophila genes

Type of gene	Locus	Reference
RNA:		
ribosomal RNA	bobbed (bb)	Tautz et al., 1988
tRNA	$tRNA^{Tyr}$	Kubli et al., 1988
enzymes	alcohol dehy-drogenase (Adh)	Goldberg, 1980 Benyajati et al., 1981
morphological:		
eye color	vermilion (v)	Pastink et al., 1989
bristle absence	achaete (ac)	Campuzano et al., 1985
developmental:		
segmentation	fushi tarazu (ftz)	Laughon & Scott, 1984
	engrailed (en)	Fjose et al., 1985
polarity	snail (sna)	Boulay et al., 1987
behavioral:		
rhythmicity	period (per)	Bargiello & Young, 1984 Reddy et al., 1984
stress response	heat shock (Hsp68)	Holmgren et al., 1979
	(Hsp70)	Southgate et al., 1983
homologs to mam-malian oncogenes	wingless (wl)	Rijsewijk et al., 1987
	sevenless (svl)	Marx, 1989
transposable elements:		
DNA elements	P	O'Hare & Rubin, 1983
retrovirus-like	copia	Rubin et al., 1981
consensus sequences:		
homeobox	bithorax (bx) Antennapedia (Antp)	McGinnis et al., 1984ab
enhancers	various	Bray et al., 1988
promoters	various	Gilmour et al., 1989; Biggin & Tjian, 1989
chromosomal structures:		
telomeres		Rubin, 1978

period (per) gene. Such flies do not lay eggs at times determined by a light conditioned biological clock as normal
flies do. The mating songs produced by mutant males are also
abnormal (see review by Konopka, 1987). Zehring et al.
(1984) were able to introduce copies of the normal allele of
the per gene into embryos homozygous for the recessive abnor-
mality. The flies that eventually hatched showed normal behavior.

Studies of RNA

If the RNA product of a particular gene can be isolated from
cells, it can be used as a probe to find the chromosomal location of the gene. One strand only of the double stranded
DNA molecule is used as a template for the formation of RNA.
RNA will thus be capable of hybridizing (base-pairing) with
the DNA strand that encoded it. Radioactively labeled RNA
can be allowed to renature to denatured DNA in chromosomes in
cytological preparations which are then exposed to a photo-
graphic emulsion. After development, the autoradiographs
will show the radioactive sites and their chromosomal location can be identified by superimposing the autoradiographs
on photographs of the same cell stained to show cytological
banding. In a similar, way labeled RNA or DNA can be used as
probes for the location of complementary sequences of DNA or
RNA in specific organs of the fly. These techniques produce
beautiful pictures in which a particular gene product can be
located (see for example Fjose et al., 1985, and the front
cover of the issue of Nature in which this paper appeared).

Hybridization between RNA and its coding DNA can be visual-
ized directly by electron microscopy. Photographs of these
hybrids show the introns of genes looping out from the paired
region since they are still present in the DNA but removed
post-transcriptionally from pre-mRNA. Recent work on post-
transcriptional modification has shown that differential
splicing is an important mechanism in developmental regulation especially in sex determination (Baker, 1989).

Historically, geneticists have used naturally occurring or

induced changes in genes (mutations) to find out what the
gene was actually doing. Since mutations occur more or less
at random, studies were limited to known mutants. Techniques
for producing RNA complementary to the normal gene product,
so-called antisense RNA, now exist. Antisense RNA, because
it can pair with DNA or RNA with complementary bases, can be
used to mimic the effects of mutation by interfering with
either the transcription or translation of the genetic material (see Weintraub, 1990). This has produced the new field
of reverse genetics. It is interesting to note that it was
in Drosophila that the first demonstration of this was carried out by H. Jäckle and colleagues (Rosenberg et al., 1985)
who introduced antisense RNA to the Krüppel gene into normal
embryos. This produced phenocopies of the developmental ab-
normalities typical of Krüppel mutations. Antisense technology promises not only to aid in genetic studies, but may also
provide possibilities for new approaches to the control of
viral diseases and growth of cancerous cells.

Studies of Proteins

Proteins make up the major part of the structures of living
organisms, and those that act as enzymes control metabolism.
Proteins can be isolated from Drosophila, and variation in
protein pattern studied by gel electrophoresis can give in-
sight into the amount of variability within individuals in a
population and between different populations (see for example
Choudhary and Singh, 1987). Other proteins bind in complex
ways to DNA and control gene activity. By using protein-spe-
cific antibodies, particular proteins can be located in specific regions of cells or of the body.

Elegant studies have allowed geneticists to start with a protein of interest, and working backwards, to locate the gene.
Antibodies can sometimes be used to identify parts of proteins while they are still being synthesized on ribosomes.
The mRNA that is the product of the gene and the blueprint
for amino acid assembly can then be isolated from the ribosomes manufacturing the protein of interest. An enzyme which
makes DNA from an RNA template (reverse transcriptase) can

then be used to produce DNA which is homologous to some of
the DNA of the gene which coded for the protein. This so-
called copy DNA can serve as a probe to locate the original
gene in the genome. Once located, the gene can then be
cloned and studied.

Biochemical techniques making use of antibodies specific for
individual proteins can be used to locate these proteins in
the body of the fly. For example, it was shown that the enzyme xanthine dehydrogenase, the product of the rosy (ry)
gene, was produced only in the Malpighian tubules and fat
bodies of the larva and not in the eye imaginal disc and yet
somehow functioned in the eye of the adult fly. This was
interpreted as meaning that the protein must be transported
to the eye. Most proteins that are exported from cells have
a group of amino acids at the amino terminal end of the molecule, known as a signal sequence, which makes their secretion
possible. Xanthine dehydrogenase lacks this typical signal
sequence. However, sequencing the amino acids of the enzyme
led to the discovery of a particular sequence of amino acids
on the carboxyl terminal end which made the enzyme capable of
entering peroxisomes which presumably are involved in its
transport (Reaume et al., 1989).

Other examples are the glass (gl) gene which is somehow needed to produce normal photoreceptor cells in the eye (Moses et
al., 1989), and snail (sna), a developmental gene (Boulay et
al., 1987). The sequencing of their products' amino acids
has shown that they contain motifs that allow binding to
DNA. This allows one to speculate that they work at the DNA
level by switching on or off other genes needed for normal
differentiation.

We hope that the selected examples mentioned above will con-
vince the reader that Drosophila is playing an important role
as an experimental animal in the burgeoning field of molecular genetics. Throughout its some eighty years of use as a
model for genetic studies, this humble organism has served
science well and will no doubt continue to do so.

Textbooks and Manuals

ALBERTS, B., BRAY, D., LEWIS, J., RAFF, M., ROBERTS, K.,
    WATSON, J.D. (eds.): Molecular Biology of the Cell, 2nd
    ed. New York: Garland 1989.
ASHBURNER, M.: Drosophila: A Laboratory Manual. Cold Spring
    Harbor NY: Cold Spring Harbor Laboratory Press 1989a.
ASHBURNER, M.: Drosophila: A Laboratory Handbook. Cold
    Spring Harbor NY: Cold Spring Harbor Laboratory Press
    1989b.
BOULNOIS, G.J. (ed.): Gene Cloning and Analysis: A Labora-
    tory Guide. Oxford: Blackwell Scientific Publications
    1987.
BROWN, T.A.: Gene Cloning. An Introduction. Wokingham,
    Berks.: Van Nostrand Reinhold (UK) 1986.
DE POMERAI, D.: From Gene to Animal. An Introduction to the
    Molecular Biology of Animal Development. Cambridge:
    Cambridge University Press 1985.
FREIFELDER, D.: Molecular Biology, 2nd ed. Boston MA:
    Jones and Bartlett 1987.
GLOVER, D. (ed.): DNA Cloning. A Practical Approach. Ox-
    ford: IRL Press 1985.
KING, R.C.: Handbook of Genetics, vol. 3: Invertebrates of
    Genetic Interest. New York: Plenum Press 1975.
MANIATIS, T., Fritsch, E.F., Sambrook, J.: Molecular
    Cloning. A Laboratory Manual. Cold Spring Harbor NY:
    Cold Spring Harbor Laboratory Press 1982.
ROBERTS, D.B. (ed.): Drosophila. A Practical Approach. Ox-
    ford: IRL Press 1986.
WATSON, J.D., HOPKINS, N.H., ROBERTS, J.W., STEIRZ, J.A.,
    WEINER, A.M.: Molecular Biology of the Gene, vols. 1 and
    2. Menlo Park CA: Benjamin/Cummings Publishing Co. 1987.

References cited

BAKER, B.S.: Sex in flies: the splice of life. Nature 340,
    521-524 (1989).
BARGIELLO, T.A., YOUNG, M.W.: Molecular genetics of a bio-
    logical clock in Drosophila. Proc. Nat. Acad. Sci. (USA)
    81, 2142-2146 (1984).
BENYAJATI, C., PLACE, A.R., POWERS, D.A., SOFER, W.: Alcohol
    dehydrogenase gene of Drosophila melanogaster: Relation-
    ship of intervening sequences to functional domains in
    the protein. Proc. Nat. Acad. Sci. (USA) 78, 2717-2721
    (1981).
BIGGIN, M.D., TJIAN, R.: Transcription factors and the con-
    trol of Drosophila development. Trends Genet. 5, 377-383
    (1989).
BOULAY, J.L., DENNEFELD, C., ALBERGA, A.: The Drosophila de-
    velopmental gene snail encodes a protein with nucleic
    acid binding fingers. Nature 330, 395-398 (1987).
BRAY, S.J., JOHNSON, W.A., HIRSH, J., HEBERLEIN, U., TJIAN,
    R.: A cis-acting element and associated binding factor
    required for CNS expression of the Drosophila melano-
    gaster dopa decarboxylase gene. EMBO J. 7, 177-188
    (1988).
CAMPUZANO, S., CARRAMOLINO, L., CABRERA, C.V., RUIZ-GOMEZ,
    M., VILLARES, R., BORONAT, A., MODOLLEL, J.: Molecular

genetics of the achaete-scute gene complex in <u>Drosophila melanogaster</u>. Cell <u>40</u>, 327-338 (1985).

CHOUDHARY, M., SINGH, R.S.: A comprehensive study of genic variation in natural populations of <u>Drosophila melanogaster</u>. III. Variations in genetic structure and their causes between <u>Drosophila melanogaster</u> and its sibling species <u>Drosophila simulans</u>. Genetics <u>117</u>, 697-710 (1987).

FJOSE, A., MCGINNIS, W.J., GEHRING, W.J.: Isolation of a homeobox-containing gene from the engrailed region of Drosophila and the spatial distribution of its transcripts. Nature <u>313</u>, 284-289 (1985).

GEHRING, W.J., HIROMI, Y.: Homeotic genes and the homeobox. Ann. Rev. Genet. <u>20</u>, 147-173 (1986).

GILMOUR, D.S., THOMAS, G.H., ELGIN, S.C.R.: Drosophila nuclear proteins bind to regions of alternating C and T residues in gene promoters. Science <u>245</u>, 1487-1490 (1989).

GOLDBERG, D.A.: Isolation and partial characterization of the Drosophila alcohol dehydrogenase gene. Proc. Nat. Acad. Sci. (USA) <u>77</u>, 5794-5798 (1980).

HOLMGREN, R., LIVAK, K., MORIMOTO, R., FREUND, R., MESELSON, M.: Studies of cloned sequences from four Drosophila heat shock loci. Cell <u>18</u>, 1359-1370 (1979).

KONOPKA, R.J.: Genetics of biological rhythms in Drosophila. Ann. Rev. Genet. <u>21</u>, 227-236 (1987).

KRISTOFFERSON, D.: The BIONET electronic network. Nature <u>324</u>, 555-556 (1987).

KUBLI, E., SUTER, B., DOERIG, R., CHOFFAT, Y.: Suppressors, introns and stage specific expression of tRNA^{Tyr} genes in <u>Drosophila melanogaster</u>. NATO Adv. Study Inst. Ser. H Cell Biol. <u>14</u>, 235-247 (1988).

LAUGHON, A., SCOTT, M.P.: Sequence of a Drosophila segmentation gene: protein structure homology with DNA-binding proteins. Nature <u>310</u>, 25-31 (1984).

MARX, J.L.: Developmental control gene sequenced. Science <u>236</u>, 26-27 (1989).

MAXAM, A.M., GILBERT, W.: A new method for sequencing DNA. Proc. Nat. Acad. Sci. (USA) <u>74</u>, 560-564 (1977).

MCGINNIS, W., LEVINE, M.S., HAFEN, E., KUROIWA, A., GEHRING, W.J.: A conserved DNA sequence in homoeotic genes of the <u>Drosophila melanogaster</u> Antennapedia and bithorax complexes. Nature <u>308</u>, 428-433 (1984a).

MCGINNIS. W., GARBER, R.L., WIRZ, J., KUROIWA, A., GEHRING, W.J.: A homologous protein coding sequence in Drosophila homoetic genes and its conservation in other metazoans. Cell <u>38</u>, 403-409 (1984b).

MOSES, K., ELLIS, M.C., RUBIN, G.M.: The glass gene encodes a zinc-finger protein required by Drosophila photoreceptor cells. Nature <u>340</u>, 531-536 (1989).

NIKI, Y., CHIGUSA, S.I., MATSUURA, E.T.: Complete replacement of mitochondrial DNA in Drosophila. Nature <u>341</u>, 551-552 (1989).

O'HARE, K., RUBIN, G.M.: Structures of P transposable elements and their sites of insertion and excision in the <u>Drosophila melanogaster</u> genome. Cell <u>34</u>, 25-35 (1983).

PASTINK, A., VREEKEN, C., NIVARD, M.J.M., SEARLES, L.L., VOGEL, E.W.: Sequence analysis of N-ethyl-N-nitrosourea-induced vermilion mutations in <u>Drosophila melanogaster</u>. Genetics <u>123</u>, 123-129 (1989).

REAUME, A.G., CLARK, S.H., CHOVNICK, A.: Xanthine dehydrogenase is transported to the Drosophila eye. Genetics 123, 503-509 (1989).
REDDY, P., ZEHRING, W.A., WHEELER, D.A., PIRROTTA, V., HADFIELD, C., HALL, J.C., ROSBASH, M.: Molecular analysis of the period locus in Drosophila melanogaster and identification of a transcript involved in biological rhythms. Cell 38, 701-710 (1984).
RIJSEWIJK, F., SCHUERMANN, M., WAGENAAR, E., PARREN, P., WEIGEL, D., NUSSE, R.: The Drosophila homolog of the mouse mammary oncogene int-1 is identical to the segment polarity gene wingless. Cell 50, 649-658 (1987).
ROSENBERG, U.B., PREISS, A., SEIFERT, E., JAECKLE, H., KNIPPLE, D.C.: Production of phenocopies by Krüppel antisense RNA injection into Drosophila embryos. Nature 313, 703-706 (1985).
RUBIN, G.: Isolation of a telomeric DNA sequence from Drosophila melanogaster. Cold Spring Harbor Symp. Quant. Biol. 42, 1041-1046 (1978).
RUBIN, G.M., BROREIN, W.J., jr., DUNSMUIR, P., FLAVEL, A.J., LEVIS, R., STROBEL, E., TOOLE, J.J., YOUNG, E.: Copia-like transposable elements in the Drosophila genome. Cold Spring Harbor Symp. Quant. Biol. 45, 619-628 (1981).
RUBIN, G.M., SPRADLING, A.C.: Vectors for P element-mediated gene transfer in Drosophila. Nucleic Acid Res. 11, 6341-6351 (1983).
SAIKI, R.K., GELFAND, D.H., STOFFEL, S., SCHARF, S.J., HIGUCHI, R., HORN, G.T., MULLIS, K.B., ERLICH, H.A.: Primer-directed enzymatic amplification of DNA with a thermostable DNA polymerase. Science 239, 487-491 (1988).
SANGER, F., NICKLEN, S., COULSON, A.R.: DNA sequencing with chain-terminating inhibitors. Proc. Nat. Acad. Sci. (USA) 74, 5463-54.. (1977).
SOUTHGATE, R., AYME, A., VOELLMY, R.: Nucleotide sequence analysis of the Drosophila small heat shock gene cluster at locus 67B. J. Mol. Biol. 165, 35-37 (1983).
TAUTZ, D., HANCOCK, J.M., WEBB, D.A., TAUTZ, C., DOVER, G.A.: Complete sequences of the rRNA genes of Drosophila melanogaster. Mol. Biol. Evol. 5, 366-376 (1988).
TSUBOTA, S.I., ROSENBERG, D., SZOSTAK, H., RUBIN, D., SCHEDL, P.: The cloning of the Bar region and the B breakpoint in Drosophila melanogaster: Evidence for a transposon-induced rearrangement. Genetics 122, 881-890 (1989).
WEINTRAUB, H.M.: Antisense RNA and DNA. Sci. Amer. 262(1), 34-40 (1990).
ZEHRING, W.A., WHEELER, D.A., REDDY, P., KONOPKA, R.J., KYRIACOU, C.P., ROSBASH, M., HALL, J.C.: P-element transformation with period locus DNA restores rhythmicity to mutant, arrhythmic Drosophila melanogaster. Cell 39, 369-376 (1984).

9. Results and Answers

1. General

1.3 Genetic Terminology

The expected phenotype is normal or wild type.

3. Transmission Genetics

3.1 Dihybrid Cross with Independent Assortment

P: Phenotype: ♀ with vestigial wings
                  ♂ with ebony body color

F$_1$:

	X; vg; +; +	Phenotype:
X; +; e; +	X vg + + X + e +	♀, normal wings normal body color
Y; +; e; +	X vg + + Y + e +	♂, normal wings normal body color

Interbreeding of F$_1$ (only autosomes carrying markers shown):

vg$^+$/vg ; e$^+$/e   x   vg$^+$/vg ; e$^+$/e

Both sexes produce the following 4 kinds of gametes:

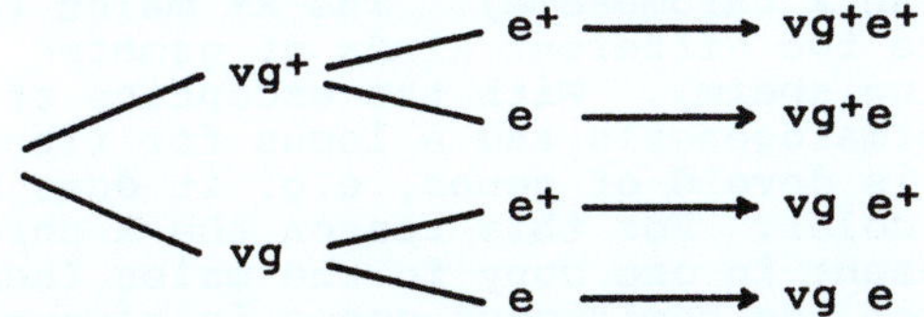

F$_2$: The progeny consist of 4 different phenotypes:

Genotype homo- zygous for		vg	e	vg e
Phenot. Wing s. Body c.	normal normal	vestigial normal	normal ebony	vestigial ebony
Expected ratio	9	3	3	1
Observed (o) Total: 1538	920	263	271	84
Expected (e)	865.1	288.4	288.4	96.1
(o-e)	54.9	−25.4	−17.4	−12.1
(o-e)^2	3014.0	645.2	302.8	146.4
(o-e)^2/e	3.5	2.2	1.1	1.5

Chi^2 = S [(o-e)^2/e] = <u>8.3</u>   0.05 > P > 0.01

<u>Answers to questions</u>
(a) P flies must be discarded from the vials because otherwise it would be possible to find mutant flies (parental phenotypes) among the F_1 and matings could occur between parents and F_1.
(b) Nonvirgin females may be used because a brother-sister
cross is performed, and all flies in the vial have the same
genotype.
(c) Nine different genotypes are expected.
(d) The ratio expected for the phenotypes is: 9 : 3 : 3 : 1.
(e) All four phenotypic classes are found in the F_2.
(f) The Chi^2 calculation leads to a P value between 0.05
and 0.01 for the probability that the difference between experiment and expectation is due to chance alone. This means
that the observed frequencies are statistically <u>not</u> in agreement with the expectation. Therefore, are Mendel's laws
wrong? No, because the observed deviations show that the
phenotypic mutant classes are always underrepresented (o-e is
negative). This is most probably due to a delayed development of vestigial and ebony flies compared with wild type
flies. For this reason it is important to collect and count
all the flies, including the late-hatching ones.

<u>3.2 Sex-Linked Inheritance</u>

	<u>Cross A</u>	<u>Cross B</u>
<u>Phenotype</u>:	♀, red eyes	♀, red eyes
	♂, red eyes	♂, <u>white</u> eyes

The XX females (homogametic sex) produce only one kind of
gamete (eggs carrying an X chromosome). The XY males (heterogametic sex) produce two different kinds of gametes
(X-bearing and Y-bearing sperm). With the exception of fertility factors for spermatogenesis and a locus for ribosomal
RNA, the Y chromosome is devoid of genes, e.g. it does not
contain genes for eye color. For this reason the X chromosomal genes are only present in one copy in the males (hemizygous). The phenotype of the hemizygous genes is always expressed according to the allele present irrespective of its
being a dominant (+) or recessive (w) one.
Note: The X chromosome in sons is always inherited from the
mother. The father passes the X chromosome on to daughters
and the Y chromosome exclusively to sons.

<u>3.3 Dihybrid Test Cross with Linked Genes</u>

P: Phenotype: ♀ with white eyes and cut wings
                   ♂ wild type

F_1:	X, w ct	Phenotype:
X, + +	X w ct X + +	♀, normal eyes normal wings
Y	X w ct Y	♂, white eyes cut wings

Test or back cross:
♀ w ct / + +   x   ♂ w ct / Y

Female gametes	Without recombination		With recombination	
	+ +	w ct	w +	+ ct
Genotype of F_2	+ + w ct	w ct w ct	w + w ct	+ ct w ct
Pheno- Eye type: Wing	normal normal	white cut	white normal	normal cut
Expected ratio	<0.5	<0.5	>0	>0
Fly Count Total = 1504	670	593	135	106

$$\frac{\text{Number of flies with recombination}}{\text{Total number of flies counted}} = \frac{135 + 106}{1504} = 0.16$$

The two genes w and ct are linked (progeny classes without
recombination are larger than those with recombination).
From the fly count above, the distance between the two genes
is estimated as 100 x 0.16 = 16 centimorgans. According to
published data the gene w maps at 1.5 and ct at 20.0 on the X
chromosome; therefore the distance is 18.5 centimorgans. The
experimentally determined values are usually underestimates
when only two markers are used.

3.4 Sex Determination

The results of the 3 crosses are (autosomes not included):

(a) ♀ X, w / X, w   x   ♂ $\overline{XY}$, w^+ / 0

    gametes of ♀ :   X, w
    gametes of ♂ : $\overline{XY}$, w^+ and   0

F_1:

	X, w
$\overline{XY}$, w^+	X, w/$\overline{XY}$, w^+
0	X, w/0

Phenotype:

$X\overline{XY}$ ♀, normal eyes
I = 1.0
X0 ♂, white eyes
I = 0.5

(b) ♀ X, w / X, w   x   ♂ X, w^+ / Y

    gametes of ♀ :   normal: X, w
                  nondisjunction: X, w/X, w   and   0
    gametes of ♂ :   normal: X, w^+ and   Y
      (nondisjunction very rare)

F$_1$:

	normal	nondisjunction	
	X, w	X, w/X, w	0
X, w$^+$	X, w/X, w$^+$ 1	X, w/X, w/X, w$^+$ 3	X, w$^+$/0 5
Y	X, w/Y 2	X, w/X, w/Y 4	Y / 0 6

Genotype	I	Phenotype	
1	1.0	♀	normal eyes
2	0.5	♂	white eyes
3	1.5	meta ♀	normal eyes, usually lethal
4	1.0	♀ with extra Y	white eyes
5	0.5	X/0 ♂	normal eyes
6	0	Y/0	lethal

(c)   ♀ X, w / X, w / Y   x   ♂ X, w$^+$ / Y

  gametes of ♀ :   X, w / X, w   and   Y   as well as
                      X, w / Y   and   X, w
  gametes of ♂ :   X, w$^+$ and   Y

F$_1$:

	X,w/X,w	Y	X,w/Y	X,w
X,w$^+$	X,w/X,w/X,w$^+$ 1	X,w$^+$/Y 3	X,w/X,w$^+$/Y 5	X,w/X,w$^+$ 7
Y	X,w/X,w/Y 2	Y/Y 4	X,w/Y/Y 6	X,w/Y 8

Genotype	I	Phenotype	
1	1.5	meta ♀	normal eyes, usually lethal
2	1.0	♀ with extra Y	white eyes
3	0.5	♂	normal eyes
4	0	Y/Y	lethal
5	0.5	♀ with extra Y	normal eyes
6	0.5	♂ with 2 Y	white eyes
7	1.0	♀	normal eyes
8	0.5	♂	white eyes

Triploid strain

In the two tables on the following page the results of the
triploid cross are given. The progeny consist of 8 aneuploid
genotypes which are lethal. In addition the meta females are
also usually lethal. 7 genotypes (a to g) lead to surviving
adults which can be distinguished phenotypically. Intersexes
are sterile due to malformation of the external and internal
sex organs. The diploid females are homozygous for dm and
therefore are sterile. In contrast, the females with an additional X chromosome are fertile and have to be selected

Checkerboard for triploid strain

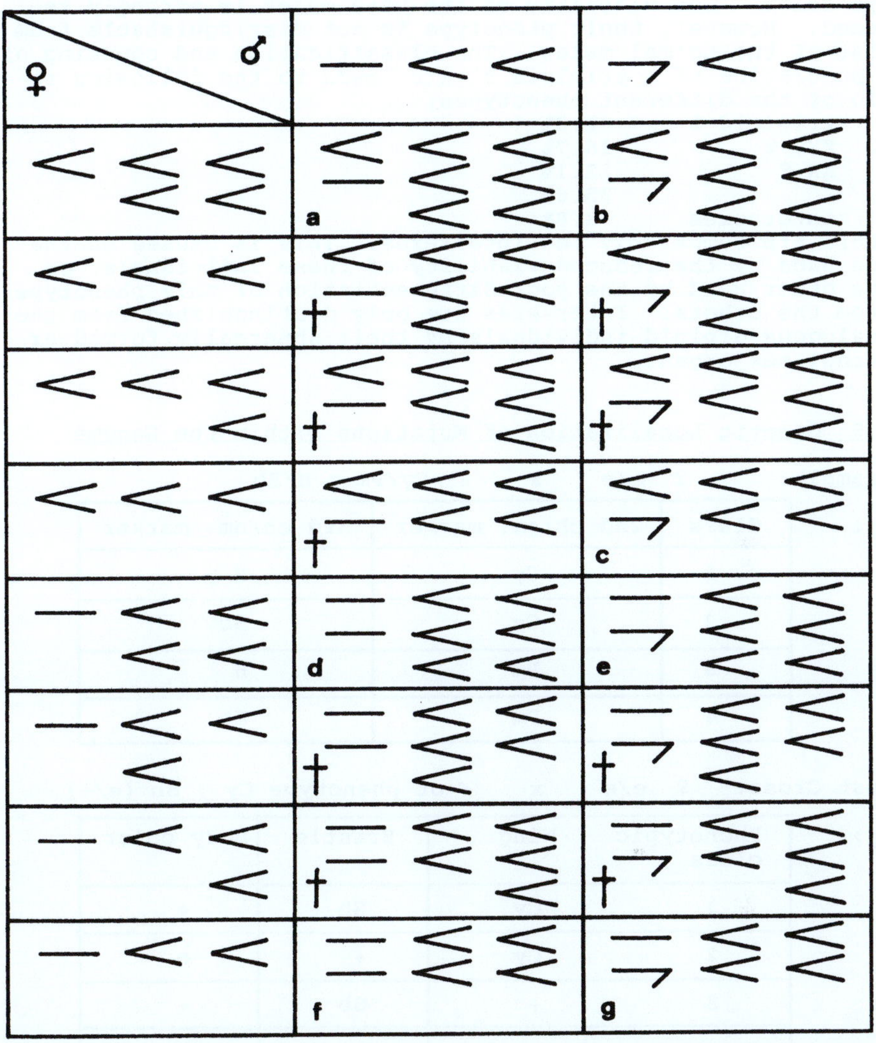

Genotype			I	Phenotype				
				Sex	Eye color	shape	Body color	Fertility
a XX/X	2/2/2	3/3/3	1	3n ♀	red	●	yellow	+
b XX/Y	2/2/2	3/3/3	0.6	intersex	orange	○	yellow	−
c XX/Y	2/2	3/3	1	♀	orange	○	yellow	+
d X/X	2/2/2	3/3/3	0.6	intersex	red	❘	yellow	−
e X/Y	2/2/2	3/3/3	0.3	meta ♂	red	❘	yellow	−
f X/X	2/2	3/3	1	♀	red	❘	yellow	−
g X/Y	2/2	3/3	0.5	♂	red	●	yellow	+

and discarded when a triploid strain is kept in continuous
culture. They are easily distinguishable due to the orange
eye color. The viability of the meta males is strongly reduced. However, their phenotype is not distinguishable from
that of the normal males. The classification and counting of
a population of a triploid strain leads to the following ratio of the different phenotypes:

triploid ♀	31.8%
XXY ♀	16.7%
XX ♀	12.1%
♂	35.6%
intersexes	3.8%

There are comparably few intersexes. This is caused on the
one hand by the reduced viability of these individuals, on
the other hand by the poor differentiation of this phenotype
from the others: Intersexes are only distinguished from the
analogous diploid individuals by their abnormally formed external sex organs.

3.5 Genetic Localization of Mutations Within the Genome

Example: ♀ e/e   x   ♂ Cy/Pm ; H/Sb

F_1:

Class	2nd chrom. marker	3rd chrom. marker
1	Cy	H
2	Cy	Sb
3	Pm	H
4	Pm	Sb

Test Cross: ♀ e/e   x   ♂ of phenotype Cy ; Sb (e/+)

F_2:

Phenotypic class	Wing	Bristle	Body color
1	Cy	Sb	+
2	Cy	+	e
3	+	Sb	+
4	+	+	e

Answers to questions

(a) ebony did not segregate from the 2nd chromosome marker Cy
(in class 2, both appear together). (b) ebony did segregate
from the 3rd chromosome marker Sb. (c) ebony assorted independently of Cy. (d) Each class has either Sb or e, never
both, never neither. (e) Therefore we conclude that e is on
chromosome 3.

3.6 Mapping on a Chromosome

Class	Phenotype			Number of flies	Type of event
	Eye	Wing	Bristle		
1	w	m	f	48	no crossover
2	+	+	+	55	
3	w	+	+	30	single c.o. w – m
4	+	m	f	26	
5	w	m	+	14	single c.o. m – f
6	+	+	f	15	
7	w	+	f	5	double c.o.
8	+	m	+	7	
			Total:	200	

$$\text{w} - \text{m} : \quad \frac{30 + 26 + 5 + 7}{200} = \frac{68}{200} = 0.340 = 34.0\%$$

$$\text{m} - \text{f} : \quad \frac{14 + 15 + 5 + 7}{200} = \frac{41}{200} = 0.205 = 20.5\%$$

Map:

$$\text{w} \longleftarrow 34 \longrightarrow \text{m} \longleftarrow 20.5 \longrightarrow \text{f}$$

The map positions of the three genes on the X chromosome
are: w = 1.5, m = 36.1, f = 56.7.

3.7 Segregation of Compound Chromosomes

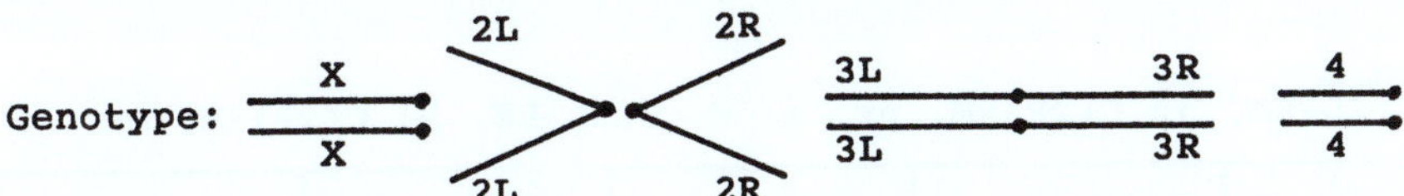

Genotype:

Expected oocytes:

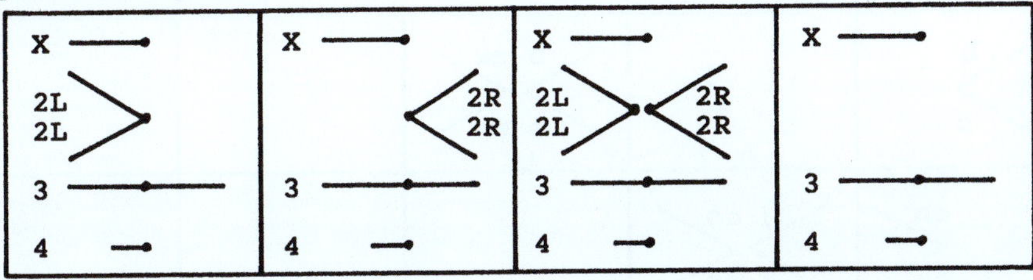

Reasons for this expectation: Usually the homologous chromosomes produce a tetrad in meiosis consisting of 4 chromatids
with 4 centromeres. In the compound chromosomes, the identical arms pair so that a tetrad is formed with four chromosome
arms but only two centromeres. If the assortment of 2L is
independent of 2R, then the 4 types of gametes given above
are expected.

Cross 1:
♀ C(2L)RM, j; C(2R)RM, px   x   ♂ wild type

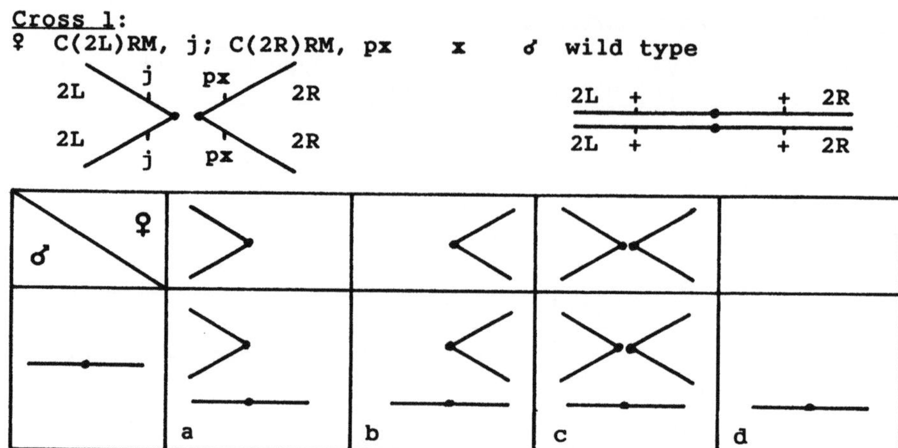

a	lethal,	2L present three times and 2R only once
b	lethal,	2L present once and 2R three times
c	lethal,	2L and 2R present three times (trisomy)
d	lethal,	2L and 2R present only once (monosomy)

Rare exceptions (approx. 1 in 10 000 zygotes) can be the result of nondisjunction of the 2nd chromosomes in spermatogenesis. Sperm which carry no 2nd chromosome or two of them, can produce viable zygotes in combination with type c or type d of the female gametes given above.

Cross 2:
♀ C(2L)RM, j; C(2R)RM, px   x   ♂ C(2L)RM, b; C(2R)RM, cn

♂ \ ♀	j / j	px / px	j px / j px	
b / b		b px / b px		
cn / cn	j cn / j cn			
b cn / b cn				b cn / b cn
			j px / j px	

<u>Cross 3:</u>
♀ C(2L)RM, b; C(2R)RM, px x ♂ C(2L)RM, j; C(2R)RM, cn

♂ \ ♀	b / b	px / px	b px / b px	
j / j		j px / j px		
cn / cn	b cn / b cn			
j cn / j cn				j cn / j cn
			b px / b px	

Table

Cross	Phenotype:	+ +	j px	j cn	b px	b cn
1	expected	0	0	0	0	0
1	observed	0	0	0	0	0
2	expected		25%	25%	25%	25%
2	observed		8	286	346	5
3	expected		25%	25%	25%	25%
3	observed		781	6	18	873

The results of crosses 2 and 3 do not agree with the expectation. Obviously, in both crosses those progeny types predominate which have inherited one compound chromosome each from both parents. From this we can conclude that in one or in both sexes the segregation of the compound chromosomes is not random. It has been shown that the segregation is random in the males, but in the females the majority of the oocytes produced contain one compound chromosome each.

<u>Note for instructor</u>: In cross 1 compound females were used because these have to be collected for cross 2. The same result is also obtained when a reciprocal cross of wild type females with C(2L)RM, j; C(2R)RM, px males is made.

3.8 Meiotic Mutants

Problem 1: Influence of the mei-9 mutation on nondisjunction frequencies

Cross A: ♀ wild type   x   ♂ X, B / Y

F_1:

	X	nondisjunction	
♂ \ ♀	X	X / X	–
X, B	X / X, B 1	X / X / X, B 3	X, B 5
Y	X / Y 2	X / X / Y 4	Y 6

Phenotype:

1	♀	XX	kidney-shaped B/+ eyes
2	♂	XY	normal eyes
3	♀	XXX	meta female, usually lethal
4	♀	XXY	normal eyes
5	♂	X0	Bar eyes
6	–	Y0	lethal

Cross B: ♀ y w^a mei-9^a/y w^a mei-9^a x   ♂ X, B / Y

F_1:

	y w^a mei-9^a	nondisjunction	
♂ \ ♀	y w^a mei-9^a	y w^a mei-9^a y w^a mei-9^a	–
X, B	y w^a mei-9^a X, B 1	y w^a mei-9^a y w^a mei-9^a X, B 3	X, B 5
Y	y w^a mei-9^a Y 2	y w^a mei-9^a y w^a mei-9^a Y 4	Y 6

Phenotype:

1	♀	XX	kidney-shaped B/+ eyes
2	♂	XY	yellow body, apricot eyes
3	♀	XXX	meta female, usually lethal
4	♀	XXY	yellow body, apricot eyes
5	♂	X0	Bar eyes
6	–	Y0	lethal

Fly Count:

Series	XX ♀	XY ♂	XXY ♀	X0 ♂	Total
wild type	1175	1051	1	0	2227
mei-9	781	577	99	67	1524

```
Frequency of nondisjunction:
wild type  :    1/2227        =  0.04%
mei-9      :    (99+67)/1524  = 10.9%
```

In females which are homozygous for mei-9 the frequency of
nondisjunction of the sex chromosomes is drastically increased. Nondisjunction of the autosomes leads to inviable
aneuploid progeny.

Problem 2: Influence of the mei-9 mutation on crossing over frequencies

♂ ♀	without recombination		with recombination		
	+ +	y cv	+ cv	y +	
	+ + →	y cv →	+ cv →	y + →	
Phenotype: Body color / Crossveins	normal present	yellow missing	normal missing	yellow present	
Fly Count: wild type / mei-9	736 / 269	429 / 165	69 / 14	61 / 8	Total: 1295 / 456

```
Frequency of recombination:
wild type  :    (69+61)/1295 = 10.04%
mei-9      :    (14+ 8)/ 456 =  4.82%
```

In homozygous mei-9 females the frequency of crossing over is
reduced. Due to the fact that in the same type of females
the frequency of nondisjunction is increased (see problem 1),
these experiments show that there is a link between crossing
over and nondisjunction in meiosis. Apparently, the chiasmata which are a consequence of crossing over, play an important (but so far not completely understood) role in the control of the correct segregation of chromatids in the two meiotic divisions. In somatic cells of mei-9 mutants a defect
in the enzymatic repair of DNA damage was found: The effects
of the mei-9 mutation in somatic cells (and most probably
also in meiotic cells) are caused by a nearly complete lack
of DNA excision repair.

4. Phenogenetics

4.1 Temperature Effect on Expression of Phenotype

```
Classification of wings:
 Series  Temp.      I   II  III   IV    V    VI  VII      N
 A       25°C       0    2   16    64  131  377  194    784
 B       29.5°C     6  345  250   134   78  105   24    942
```

216

$$N_A = 784 \qquad\qquad N_B = 942$$
$$Z(f_i x_i) = 4583 \qquad Z(f_i x_i) = 3170$$
$$\bar{x}_A = 5.85 \qquad\qquad \bar{x}_B = 3.37$$

$$S = \sqrt{1.6} = 1.26$$
$$t = 40.7$$
$$d.f. = 1724$$
$$P \ll 0.05$$

The difference between the distributions of the phenotypic
classes found in the two series at 25°C and 29.5°C is significant. At the higher temperature the wings of the vg/vg
flies are significantly longer than at the normal culture
temperature. The development of the wing is under the control of a complex polygenic system including the gene vestigial among many others. By increasing the culture temperature many temperature-dependent reaction constants are affected in various ways, and in consequence of this a longer
wing is produced in the vg/vg genotype. When culture temperature is increased, the penetrance of the genotype becomes
incomplete. It sinks from 784/784 = 100% to 936/942 =
99.4%. For judging the expressivity, we only consider the
classes in which the vestigial phenotype is manifested, i.e.
we leave out class I. Only 6 wings in the 29.5°C series have
to be deleted as compared with the calculation of the mean
given above. The mean for the expressivity at 29.5°C is
3164/936 = 3.38. This value has to be compared with the mean
at 25°C (with 100% penetrance) of 5.85. This value calculated in relation to the class numbering shows that the wings
are substantially smaller at 25°C.

4.2 Temperature Effect on a Homeotic Mutant

The classification of ss^{a40a} flies raised at different culture temperatures gave the following results:

Temp.	Number of flies	Both antennae wild type (%)	Intermediate (%)	Both antennae transformed (%)
18°C	110	0	35	65
20°C	112	32	36	32
25°C	99	84	16	0
29°C	110	0	13	87

The higher culture temperature affects both penetrance and
expressivity of this gene: The penetrance at 25°C is only
16% whereas it is 100% at 29°C. In addition the expressivity
is higher at the higher temperature, i.e. there are more
strongly transformed phenotypes. The product of this gene
behaves most like the wild type product at a temperature of
25°C. From the additional data given above with results from
lower temperatures it can be seen that the effect of temperature is not continuous because at 18°C the penetrance and
expressivity are again increased compared to 25°C.

4.3 Mutants with Abnormal Eye Color

Answers to questions
(a) In bw/bw animals no red eye pigments are produced.
(b) In st/st flies the brown eye pigments are missing.
(c) In the F_2 a 9 : 3 : 3 : 1 ratio of wild type : brown : scarlet : white eyes is expected.

F_1:

♂ \ ♀	bw st⁺
bw⁺ st	bw st⁺ / bw⁺ st

normal eye color

All F_1 individuals show a normal wild type eye color.

F_2: ♀ bw⁺/bw; st⁺/st   x   ♂ bw⁺/bw; st⁺/st

♂ \ ♀	bw⁺ st⁺	bw st⁺	bw⁺ st	bw st
bw⁺ st⁺	bw⁺ st⁺ / bw⁺ st⁺ normal	bw st⁺ / bw⁺ st⁺ normal	bw⁺ st / bw⁺ st⁺ normal	bw st / bw⁺ st⁺ normal
bw st⁺	bw⁺ st⁺ / bw st⁺ normal	bw st⁺ / bw st⁺ brown	bw⁺ st / bw st⁺ normal	bw st / bw st⁺ brown
bw⁺ st	bw⁺ st⁺ / bw⁺ st normal	bw st⁺ / bw⁺ st normal	bw⁺ st / bw⁺ st bright red	bw st / bw⁺ st bright red
bw st	bw⁺ st⁺ / bw st normal	bw st⁺ / bw st brown	bw⁺ st / bw st bright red	bw st / bw st white

In the F_2 a double homozygous class, bw/bw; st/st, is expected. This corresponds to the white (or slightly yellowish) eye color. The very light color shows that both pigment components are missing in these individuals. Therefore, this experiment confirms the hypothesis that the brick red eye color of the wild type is produced by two different types of pigment.

Notes for instructor
(1) The yellowish color in bw/bw; st/st flies is the result of a block in the synthesis of the brown pigments in the mutant st in a step where yellowish intermediate products are accumulated.
(2) For technical reasons, in this experiment it is advisable to use the st pᴾ/st pᴾ strain instead of the st/st strain (see also experiment 5.2). The addition of the gene pᴾ leads to a completely white eye color in the bw/bw; st pᴾ/st pᴾ individuals. The genes st and pᴾ are closely linked so that there is hardly any crossing over in the heterozygous F_1 females, and therefore the phenotype pᴾ (dull red) appears only very rarely. The gene pᴾ has been omitted on purpose in order to illustrate the fundamental principle of the experiment.
(3) The experiment can also be done with the reciprocal cross.

4.4 Chromatographic Analysis of Eye Color Mutants

The evaluation of the chromatogram gives the following inventory of UV fluorescent pterines in the 5 different mutants:

Strain:	+	se	ry^2	w^a	w
Isosepiapterin	+	+	+	(−)	−
Biopterin	+	++	++	(−)	−
2-Amino-4-hydroxypteridin	+	++	++	(−)	−
Sepiapterin	+	+++	+	(−)	−
Xanthopterin	+	++	+	(−)	−
Isoxanthopterin	+	++	−	(−)	−
Drosopterins	+	−	(+)	(−)	−

<u>sepia</u>: Sepiapterin, isoxanthopterin and xanthopterin are increased, the drosopterins are missing. Block between sepiapterin and drosopterins.
<u>rosy</u>: 2-Amino-4-hydroxypteridin increased. Isoxanthopterin missing. Hypothesis: Deficiency of the xanthine dehydrogenase activity. Can also be shown enzymatically. Traces of drosopterins present. Probably a second, less efficient synthesis pathway for the drosopterins via xanthopterin present.
<u>white-apricot</u>: All pigments are in general strongly reduced. Hypothesis: Probably a defect present in the binding of the pigments to the protein granules (which are synthesized under the control of the w^+ locus) so that the synthesized pigments are not stabilized and not maintained.
<u>white</u>: No pigments whatsoever are found. Hypothesis: Lack of the protein granules required for the binding of the pigments.

4.5 Transplantation of Eye Imaginal Discs

The synthesis pathway including the intermediate products (as determined biochemically) is as follows:

$$\text{tryptophan} \xrightarrow{v^+} \text{formyl-kynurenine} \longrightarrow \text{kynurenine} \xrightarrow{cn^+} \text{hydroxy kynurenine} \xrightarrow{st^+} \text{brown eye pigments}$$

The correct temporal sequence of the gene functions therefore is: v − cn − st. It is evident that at the most, as many synthesis steps can be distinguished as there are mutants available. Those steps in which no mutation of the gene responsible is available cannot be identified. In addition, in a larger series of various mutants it is conceivable that several mutants affect the same step.

<u>Additional problems</u>
(a) The +/+ eye disc is fully autonomous in producing all the wild type pigments.
(b) Most probably the intermediate product(s) required in a st/st disc for the production of the brown pigments is not able to diffuse into it and, therefore, is not able to cause the synthesis of these pigments.

4.6 Supplementation of an Eye Color Mutant

We find the following eye colors of the flies in the four
different series:

v; bw	+	antibiotics :	white eyes
v; bw	+	kynurenine :	brown eyes
cn bw	+	antibiotics :	white eyes
cn bw	+	kynurenine :	white eyes

We see that in both strains the expected white eye color is
present in the vials treated with antibiotics only. However,
if kynurenine is fed, brown pigments are produced in the
strain v; bw but not in the strain cn bw. This shows that in
the mutant vermilion the block in the synthesis pathway can
be circumvented by feeding kynurenine, while this is not possible in the mutant cinnabar which is blocked later. (Compare also with solution of experiment 4.5).

Note for instructor: Should the kynurenine not dissolve completely, it can also be fed as a suspension. The experiment
can be continued for another generation by intercrossing the
brown-eyed v; bw flies which have been treated with kynurenine. Their progeny will show again the white eye color which
demonstrates that the genotype of the flies has not been altered by the kynurenine treatment.

4.7 Genetic Complementation and Allelism

Example:

Phenotype of heterozygote	Conclusion
wild type	nonallelic
mutant	allelic

4.8 Transposable Elements

The analysis of the ovaries leads to the following results:

Cross	Number of females with:			Dysgenic ovaries (%)
	Both ovaries normal	One ovary normal	No normal ovaries	
1	20	0	0	0
2	20	0	0	0
3	20	0	0	0
4	0	1	19	39/40 =97.5%

Dysgenesis is only observed when P elements are introduced
into an M cytotype at high temperature (cross 4).

Other possible control crosses:

♀	w(M)	x	♂	w(M)	at	20°	and	28°C
♀	w$^+$(P)	x	♂	w$^+$(P)	at	20°	and	28°C

5. Mutation Genetics

5.1 Induction and Detection of Sex-Linked Recessive Lethals

P: ♀ Basc, w^a B / Basc, w^a B   x   ♂ X / Y
                                                    EMS-treated

♂ ＼ ♀	Basc, w^a B	Phenotype:
X(EMS)	Basc, w^a B X(EMS)	kidney-shaped B/+ eyes, normal eye color, ♀
Y	Basc, w^a B Y	narrow bar eyes, apricot eye color, ♂

F_1:   ♀ Basc, w^a B / X(EMS) x ♂ Basc, w^a B / Y

♂ ＼ ♀	Basc, w^a B	X(EMS)
Basc, w^a B	Basc, w^a B Basc, w^a B	X(EMS) Basc, w^a B
Y	Basc, w^a B Y	X(EMS) Y

F_2:

Genotype	Phenotype		
	Sex	Eye shape	Eye color
Basc,w^a B/Basc,w^a B	♀	Bar	apricot
Basc,w^a B/X	♀	kidney	normal (brick red)
Basc,w^a B/Y	♂	Bar	apricot
X/Y	♂	round	normal (brick red)

In the F_2 the EMS-treated X chromosome is hemizygous in all
X/Y males. A recessive lethal induced in this chromosome
will manifest itself phenotypically, i.e. it will lead to the
lack of this Mendelian class in the F_2. The parents need
not be discarded because the male has an apricot Bar eye
which does not interfere with the classification of a lethal. The F_1 females must be cultured individually because
each of them represents one single X chromosome originating
from one treated sperm. In each F_2 vial only <u>one</u> X chromosome can be tested for the presence of a recessive lethal.
When only about 100 tests are analyzed in each series the
following results are obtained:

	<u>Without lethal</u>	<u>With lethal</u>	<u>Rate</u>
Control	89	1	1/90 = 1.1%
25 mM EMS	81	11	11/92 = 12.0%

In a more extensive study with EMS treatment the following
results were obtained: 0.1 mM: 42/6296 = 0.67%; 0.5 mM:
113/5549 = 2.0%; 1.0 mM: 102/2540 = 4.0%; 25 mM: 1028/2390
= 43%.

<u>Note for instructor:</u> In the long-term culture of the Basc
strain flies with normal round eyes occur occasionally. This
is caused by rare unequal crossing overs within the Bar region. Bar is a duplication (Bridges, 1936). Unequal crossing over leads to a reverted Bar (normal eye size) and a
triplication (still narrower eyes than Bar) (Sturtevant,
1925). The atypical chromosomes with the reverted Bar can be
eliminated by discarding the aberrant males and females (with
B/+ phenotype). The strain itself is not affected by such a
crossing over event.

<u>Literature</u>
BRIDGES, C.B.: The Bar "gene" a duplication. Science <u>83</u>,
    210-211 (1936). In: Classic Papers in Genetics (ed.
    J.A. Peters). Englewood Cliffs NJ: Prentice-Hall 1959.
STURTEVANT, A.H.: The effects of unequal crossing over at
    the Bar locus in Drosophila. Genetics <u>10</u>, 117-147
    (1925). In: Classic Papers in Genetics (ed. J.A.
    Peters). Englewood Cliffs NJ: Prentice-Hall 1959.

5.2 <u>Induction and Detection of Reciprocal Translocations</u>

P: ♀ X/X; 2*/2*; 3*/3*   x   ♂ X/Y; 2/2; 3/3
                                   (irradiated)

♀ \ ♂	X ; 2 ; 3	Y ; 2 ; 3
X ; 2* ; 3*	X/X ; 2/2* ; 3/3*	X/Y ; 2/2* ; 3/3*

All F_1 individuals, females (X/X) as well as males (X/Y),
are heterozygous for the recessive markers (2*/2; 3*/3 =
bw/bw+; st p^P/st+ p^{P+}) and therefore have normal red
eyes.

F_1: ♀ X/X; 2*/2*; 3*/3*   x   ♂ X/Y; 2/2*; 3/3*

Class:	A	B	C	D	E	F	G	H
	X23	Y23	X2*3	Y2*3	X23*	Y23*	X2*3*	Y2*3*
X2*3*	X/X 2*/2 3*/3	X/Y 2*/2 3*/3	X/X 2*/2* 3*/3	X/Y 2*/2* 3*/3	X/X 2*/2 3*/3*	X/Y 2*/2 3*/3*	X/X 2*/2* 3*/3*	X/Y 2*/2* 3*/3*
Sex	♀	♂	♀	♂	♀	♂	♀	♂
Eye color	red	red	brown	brown	yel-low	yel-low	white	white

When a 2/3 translocation is present, classes B and D become:

<u>Class B</u>

♀ \ ♂	Y 2^T 3^T
X 2* 3*	X 2* 3* Y 2^T 3^T
	viable

<u>Class D</u>

♀ \ ♂	Y 2* 3^T
X 2* 3*	X 2* 3* Y 2* 3^T
	lethal

Progeny of class B contain an euploid genome with a complete
reciprocal translocation and are, therefore, viable. Progeny
of class D are aneuploid; they contain the segment of chromo-

some 2 involved in the translocation three times, but the segment of chromosome 3 only once. Such individuals are lethal, i.e. in the F_2 the males with brown eye color are missing.

In the case of the translocations T1 ($2^T/3^T$), T2 ($Y^T/2^T$) and T3 ($Y^T/3^T$) the following classes of progeny are present in the F_2:

T1	T2	T3
female, red	male, red	male, red
male, red	female, brown	male, brown
female, white	male, yellow	female, yellow
male, white	female, white	female, white

In an actual experiment these results were obtained:

	Normal	T1	T2	T3	Total	Rate
Control	211	–	–	–	211	0/211 = 0%
4000 R X-rays	336	69	1	2	408	72/408 = 17.6%

Such a frequency corresponds to the expectation in as much as one finds approximately 20% translocations in larger experiments (Gonzales, 1972). The low rate of Y-autosome translocations is due to the fact that F_1 males which carry such a translocation are very often sterile.

5.3 Mutagen-Induced Sex Chromosome Losses

__Cross 1:__   ♀ X, y w   x   ♂ X^{C2}, y f / y^+ Y B^S

In the checkerboard below only the phenotypes are given (sex, eye shape, eye color and body color).

♂ \ ♀	normal X, y w	nondisjunction X, y w/X, y w	–
X^{C2},y f	♀ normal red yellow	lethal	lethal*
y^+Y B^S	♂ Bar white normal	♀ kidney-shaped white normal	lethal
–	X0 ♂ normal white yellow	♀ normal white yellow	lethal
y^+	♂ normal white normal	♀ normal white normal	lethal
B^S	♂ Bar white yellow	♀ kidney-shaped white yellow	lethal

* These X/0 ♂ die because the ring-X chromosome contains a recessive lethal which manifests itself only in the absence of a Y chromosome (Y-suppressed lethal).

<u>Literature on Y-suppressed lethals</u>
LINDSLEY, D.L., EDINGTON, C.W., VON HALLE, E.S.: Sex-linked recessive lethals in Drosophila whose expression is suppressed by the Y-chromosome. Genetics <u>45</u>, 1649-1670 (1960).
LINDSLEY, D.L., EDINGTON, C.W., VON HALLE, E.S.: The effect of gametic genotype on the radiation sensitivity of Drosophila sperm. In: Repair from Genetic Radiation Damage (ed. F.H. Sobels). Oxford: Pergamon Press 1963.

<u>Cross 2</u>: ♀ X, y w^a mei-9^a x ♂ X^{C2}, y f / y^+ Y B^S

In this cross essentially the same types of progeny are produced as in cross 1. Only the white eye color marker w of the first cross is replaced here by w^a, which results in apricot eyes.

<u>Fly count</u>:

	XX ♀	XY ♂	X0 ♂	PL ♂	PL ♂	ND ♀	Total (−ND)
Eye form Eye color	+ +	B w(a)	+ w(a)	B w(a)	+ w(a)	B/+/+ w(a)	
Body col.	y	+	y	y	+	+	
y w Control	244	260	3	−	−	−	507
y w MMS	1148	1211	45	−	−	4	2404
mei-9^a Control	206	192	5	−	−	−	403
mei-9^a MMS	608	936	61	1	2	5	1608

<u>Rates of chromosome losses</u>:

y w	Control :	3 / 507	= 0.59%
y w	MMS :	45 / 2404	= 1.87%
mei-9^a	Control :	5 / 403	= 1.24%
mei-9^a	MMS :	64 / 1608	= 3.98%

Both spontaneously as well as after MMS treatment, the rates of sex chromosome losses are higher with mei-9 females than with y w females which are excision repair-competent. DNA damages in the chromosomes of the sperm which are normally repaired persist in the mei-9 eggs and lead to chromosome losses in the first round of replication, i.e. in the duplication of chromosomes before the first cleavage division.

5.4 Somatic Mutation and Mitotic Recombination

Problem 1: Wing imaginal disc

♀ + flr^3 + / TM3, + + Ser   x   ♂ mwh + + / mwh + +

♂ \ ♀	+ flr +	+ + Ser
mwh + +	+ flr + mwh + +	+ + Ser mwh + +

Phenotype: normal wings   Serrate wings

The microscopic analysis of wings from adults which were irradiated as larvae with 1500 R X-rays gives these results:

Experiment	Number of wings (w)	Single spots s	s/w	Twin spots t	t/w
Control	96	21	0.22	1	0.01
X-rays	12	43	3.58	4	0.33

Both in the control as well as in the irradiated series one
finds more single spots than twin spots. Twin spots are the
result of mitotic recombination in the chromosome segment
between the centromere and the flr locus. Single spots are
the result of mitotic recombination between the flr and the
mwh locus. In addition, single spots can also be produced by
mutation or deletion. The flare clones have a reduced viability compared with the multiple wing hairs clones and can
be lost during development. In this case it is possible that
instead of a twin spot only an mwh single spot is left. The
frequency of small spots, i.e. only 1 or 2 cells in size, is
much higher among the single spots than among the twins.
These small spots are the consequence of mutational events
occurring late in development of the wing, or they represent
clones of aneuploid cells which are unable to divide further.

Problem 2: Eye imaginal disc

♀ w^{CO} sn/w^{CO} sn ; se/se   x   ♂ w sn$^+$/Y ; se/se

♂ \ ♀	w^{CO} sn ; se	Phenotype:
w sn$^+$; se	w^{CO} sn se w sn$^+$ se	♀, intermediate eye color
Y; se	w^{CO} sn se Y se	♂, dark brown eye color

The arrangement of the sectors in the eye shows that there
are groups of ommatidia which always develop in the same way
from individual cell lines during eye differentiation.

6. Population Genetics

6.1 Determination of Allele Frequencies in Populations in Hardy-Weinberg Equilibrium

Frequency of T alleles = p, frequency of t alleles = q.
Total p+q = 1.

<u>Parental generation</u>   Oocytes

		p (T)	q (t)
Sperm	p (T)	p^2 (T/T)	$p{\cdot}q$ (T/t)
	q (t)	$p{\cdot}q$ (T/t)	q^2 (t/t)

Frequencies in the F_1: p^2 (T/T) & $2pq$ (T/t) & q^2 (t/t)

Frequency of oocytes produced by the females:

	T oocytes	t oocytes
T/T females	p^2	–
T/t females	pq	pq
t/t females	–	q^2
Total:	$p^2 + pq$	$pq + q^2$
	$= p(p+q)$	$= q(p+q)$ Note: p+q=1
	$= p$	$= q$

This shows that the females of the F_1 produce T and t
oocytes in the same frequencies as the females in the
parental generation.

6.2 Variations in Populations with Natural and Artificial Selection

Experiment A: NATURAL selection
The simplest assumption is that the Hardy-Weinberg law can be
applied. We assume: (1) Panmixia (no mating preferences),
(2) equal fertility of all genotypes, (3) equal viability of
all genotypes. Therefore we expect that the population will
show the following composition in all generations:

Genotypes	A/A	A/a	a/a
Frequency	p^2	$2pq$	q^2

Phenotypes	A	a
Frequency	$p^2 + 2pq$	$q2$

This prediction should be valid especially for the F_2 because the F_2 is the progeny of many crosses A/a x A/a. Due
to p = q = 0.5, the frequencies of the phenotypes should be A
= 0.75 and a = 0.25.

The results of an experiment using the vg marker are as follows:

Gene-ration	Total flies	Number of A flies	Number of a flies	Frequency of a flies (%)
F_2	1948	1516	432	22.2
F_3	1179	1001	178	15.1
F_4	773	710	63	8.2
F_5	2190	1902	288	13.2

We see that the result is not in agreement with the expectation. Apparently one of the assumptions is not fulfilled: Not all genotypes have the same viability. The homozygous a/a flies show reduced survival and are present in lower frequencies than expected.

Experiment B: ARTIFICIAL Selection

In this experiment in each generation the homozygous a/a flies were eliminated completely. Thus our population has the following composition:

	A/A	A/a	a/a	Frequency of a
Initial population	p^2	$2pq$	q^2	q
After selection	$\dfrac{p^2}{p^2 + 2pq}$	$\dfrac{2pq}{p^2 + 2pq}$	0	$\dfrac{q}{1+q}$

The frequency of a in the population used for subculturing is therefore (note that p+q = 1):

$$\frac{1}{2} \cdot \frac{2pq}{p^2 + 2pq} = \frac{pq}{p^2 + 2pq} = \frac{pq}{p(p + 2q)} = \frac{q}{1 + q}$$

We see that the formula for the q values of two subsequent generations is:

$$q_2 = \frac{q_1}{1 + q_1}$$

This is an equation of an harmonic sequence. For this reason it is possible to give a general formula for the q values which allows to determine this value for any generation n always with complete elimination of the homozygous genotypes:

$$q_n = \frac{q_0}{1 + n.q_0}$$

Our experiment started with a population in which p = q = 0.5. From this we find with the formula above these theoretically expected q values:

Generation	q	q^2
F_2	0.5	0.25
F_3	0.33	0.108
F_4	0.25	0.062
F_5	0.20	0.040
F_6	0.166	0.027
F_7	0.142	0.020

In an actual experiment with vestigial flies the following
result was obtained:

Gene-ration	Total flies	Number of A flies	Number of a flies	Frequency of a flies (%)
F_2	1948	1516	432	22.2
F_3	1413	1245	168	11.9
F_4	1684	1575	109	6.5
F_5	1603	1457	146	9.1

A comparison of the theoretical and the experimental results
is shown in this graph:

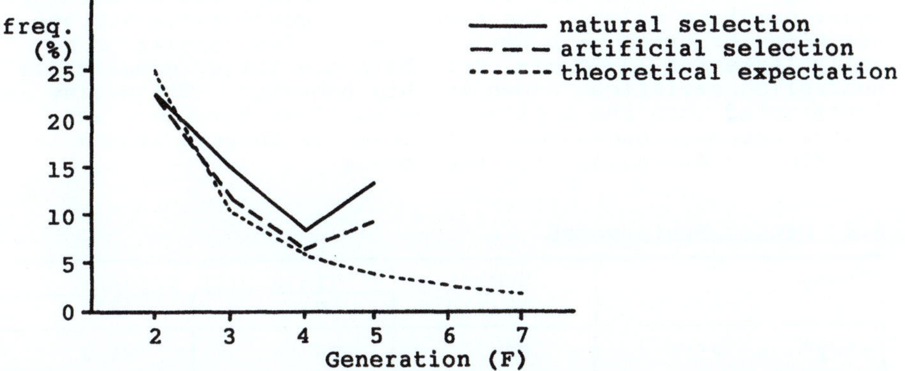

From these results it can be concluded that it is not possible to eliminate a recessive mutant completely from a population by continuous selection against the homozygote only.

6.3 Mating Behavior

In nature the two sexes of the fruit fly meet at places where
they feed, for example on fresh or also on fermenting fruit.
First the male orientates himself relative to the female:
The male is positioned with front towards the female and lat-
erally close to her. This orientation which occurs at a distance of approx. 2 to 3 mm is the first element in the mating
behavior. The male tries to keep this orientation position
later on when the female moves away. When the female is im-
mobile for a while, the male starts to circle the female with
his head always towards her. After some time the male starts
to touch the female with his forelegs. This element of the
mating behavior is not essential in _Drosophila melanogaster_
in contrast to other species. While the male is circling the
female he starts to vibrate one wing repeatedly; he spreads
the wing closer to the female at a right angle and vibrates
it rapidly before bringing it back into normal position. In
this display of the wing the male lowers the rear edge of it

so that the wing surface is inclined towards the female. Vibration phases normally last only for a few seconds, but are often repeated many times. During longer mating rituals one can observe longer and shorter vibration phases in irregular intervals. If the female does not move, the male then positions himself directly behind the female with his head close to the female's abdomen. Now the male starts to lick the genitalia of the female with his proboscis. This happens immediately before a mating attempt. The male positions himself with his forelegs under the abdomen of the female and starts to bend his abdomen under the forepart of his own body towards the female genitalia. This movement is accompanied by a mounting of the front legs over the female abdomen so that the legs help to spread the female wings. When the female is receptive for a copulation, then the wings are pushed aside with the head and the front legs. The male embraces the female abdomen with his second pair of legs and clutches himself with the claws of the front legs at the base of the female wings. The sex combs help to hold on. The copulation takes place in this position. Normally the pair remains motionless during this phase. The female may feed or ward off other courting males. The female may occasionally vibrate her wings or tread with her hind legs. The copulation normally lasts for 15-20 min, but there are large, genetically controlled variations known in this behavior. The mating is interrupted when the female tries to close her wings and pushes the male backwards. The male has to turn around 180° in order to disengage from the female.

6.4 Mating Preferences

Cross	Number of vials	Vials with progeny	
		(n)	(%)
y^+/y^+ x y^+/Y	82	78	95.1
y^+/y^+ x y/Y	85	32	37.6

The yellow males compared to wild type males have a drastically reduced rate of successful matings with wild type females. This is caused by a mutant mating behavior because the yellow males vibrate their wings in much shorter phases and with longer intervals. Apparently the females receive these stimuli with the antennae less well, which leads to the mating failure of the yellow males.

7. Cytology and Cytogenetics

7.1 Microscopic Analysis of Mitotic and Polytene Chromosomes

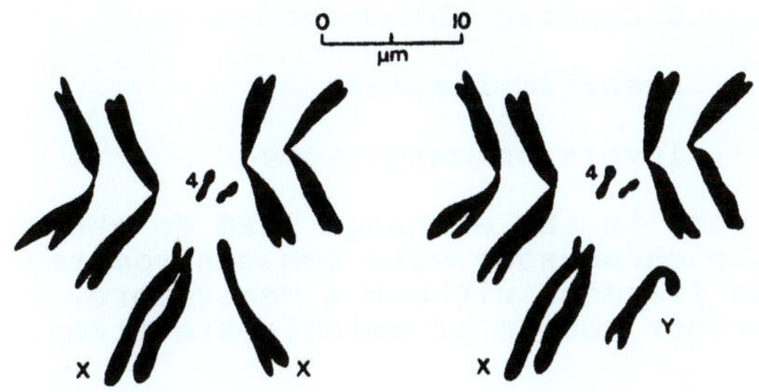

Chromosomes of dividing cells in the neural ganglion of wild type larvae. From Ransom (1982).

7.2 Balancer Chromosomes

Problem 1: Analysis of inversion loops

The black triangles and the points in the diagrams give the positions of the breaks. The analysis of these inversion loops leads to these results:

(a) Chromosome with two separate inversions:

       A▲E D C B▲F▲H G▲I

       A.B C D E.F▲H G▲I

       A.B C D E.F.G H.I

(b) Chromosome with one inversion enclosing the other:

       A B▲H G▲E F▲D C▲I

       A B▲H G.F E.D C▲I

       A B.C D.E F.G H.I

(c) Chromosome with two overlapping inversions:

       A▲F G H▲D C B▲E▲I

       A.B C D▲H G F.E▲I

       A.B C D.E.F G H.I

Problem 2: Crossing over products in inversion heterozygotes

The four chromatids have the following structure after the
crossing over event:

(1) A B C D E F   (5)   normal chromosome

(2) A B C D E A   (4)   dicentric chromosome

(6) F E D C B F   (8)   acentric fragment

(3) A E D C B F   (7)   inversion chromosome

The two chromatids taking part in the exchange lead to abnormal products: one dicentric chromosome with two centromeres
and one acentric chromosome fragment without a centromere.
The consequences in the further course of meiosis are given
in this diagram:

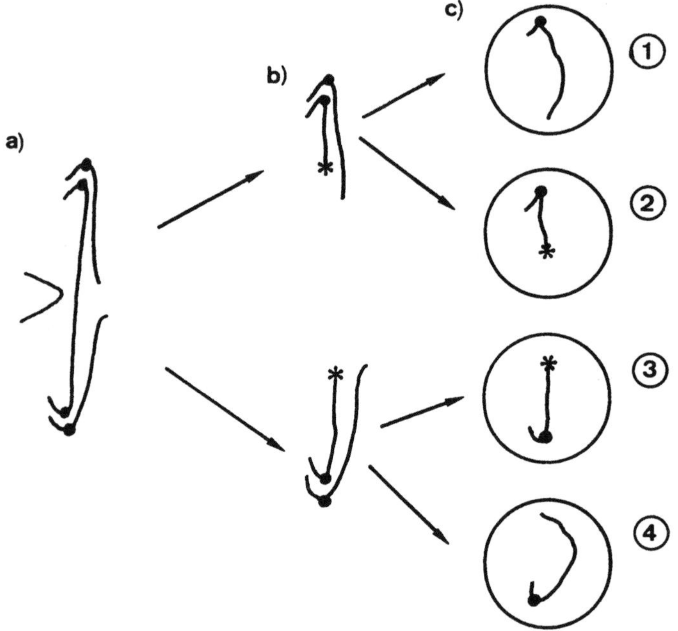

a) The dicentric chromatid leads to a chromatid bridge in
anaphase of the first meiotic division. The acentric fragment is unable to move to a spindle pole; it lags and gets
lost.
b) The chromatid bridge breaks. This happens at the physi-
cally weakest point of the chromatid which is not identical
with the crossover position.
c) In the subsequent second meiotic division four gametes
are formed. Only two of them contain the complete chromosome: No. 1 the normal chromosome and no. 4 the inversion
chromosome. The other two gametes lack the information lost
with the acentric fragment and in addition they contain a
fragment of the broken dicentric chromosome. Gametes no. 2
and 3 lead to aneuploid, inviable combinations after insemi-

nation. Furthermore they contain chromosome fragments with
open breaks which can lead to so-called breakage-fusion-
bridge cycles (for further details see Rieger et al., 1976).
In certain cases the chromosome bridge does not break. This
leads to an arrest of the meiotic division and thus to the
loss of a whole tetrad of gametes.

Based on these observations one expects that in inversion
heterozygotes the fertility should be reduced depending on
the size of the inversion. In _Drosophila melanogaster_ a very
special situation is encountered: In contrast to the expectation, inversion heterozygotes show normal fertility. For
example, in the test for sex-linked recessive lethals the
Basc/+ females are just as fertile as are the +/+ or Basc/
Basc females. Nevertheless, practically all crossover products of the X chromosome are eliminated completely by the
two inversions (one enclosing the other) of the Basc chromosome. Beadle and Sturtevant (1935) have shown that in the
female meiosis the chromosome bridges lead to a new orientation of the diads. Because the three haploid products of
meiosis lying closest to the egg surface are eliminated as
polar bodies, the chromatids involved in the crossing over
within the inversion loop are therefore also eliminated.
The figure below summarizes these events (after Strickberger,
1976).

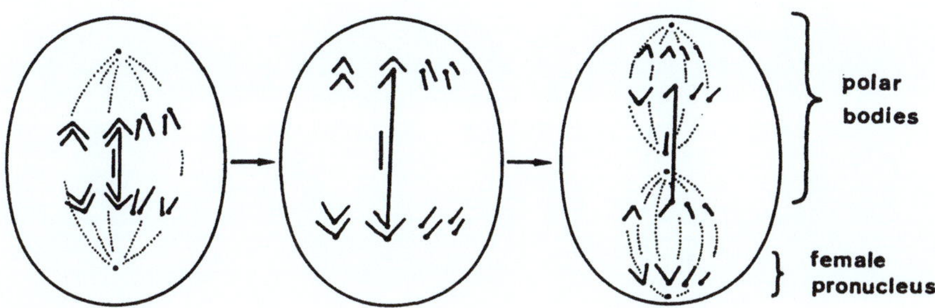

Literature
BEADLE, G.W., STURTEVANT, A.H.: X-chromosome inversions and
    meiosis in _Drosophila melanogaster_. Proc. Nat. Acad.
    Sci. (USA) 21, 384-390 (1935).
RANSOM, R. (ed.): A Handbook of Drosophila Development.
    Amsterdam: Elsevier Biomedical Press 1982.
RIEGER, R., MICHAELIS, A., GREEN, M.M.: Glossary of Genetics
    and Cytogenetics. Berlin: Springer-Verlag 1976.
STRICKBERGER, M.W.: Genetics, 2nd ed. New York: MacMillan
    Publishing Co. 1976.

10. Subject Index

MIX
Papier aus verantwortungsvollen Quellen
Paper from responsible sources
FSC® C105338

If you have any concerns about our products,
you can contact us on
ProductSafety@springernature.com

In case Publisher is established outside the EU,
the EU authorized representative is:
Springer Nature Customer Service Center GmbH
Europaplatz 3, 69115 Heidelberg, Germany

Printed by Libri Plureos GmbH
in Hamburg, Germany